T0364677
ZH-27

BMW MOTORCYCLES 100 YEARS

20
40
60
80
100
120
140
160
180
200
km
2
3
4
5
6
7

BMW MOTORCYCLES 100 YEARS

ALAN DOWDS

In memory of my dad, Jim Dowds, who passed away as I was writing this book, and who first showed me the joys of working on a motorcycle engine...

First Published in 2022 by Motorbooks, an imprint of The Quarto Group,
100 Cummings Center, Suite 265-D, Beverly, MA 01915, USA.
T (978) 282-9590 F (978) 283-2742 Quarto.com

26 25 24 23 22 1 2 3 4 5

ISBN: 978-0-7603-7471-9

Digital edition published in 2022

eISBN: 978-0-7603-7472-6

Library of Congress Cataloging-in-Publication Data available

Design: Silverglass
Jacket photos: front: Henry von Wartenberg; front flap: BMW Archive; back: BMW Archive;
back flap: John Noble/*SuperBike* magazine
Hardcover photos: BMW Archive
Endpapers: front: BMW Archive; back: BMW/James Wright
Photography: BMW Archive unless otherwise noted

Printed in China

CONTENTS

B M W
1
BMW
Michel
Offelsmeyer
Flugmotoren – Motorräder – Bootsmotoren
Bayerische Motoren Werke A.-G., München

INTRODUCTION

A hundred years is a long time for a human being—though it is, of course, the merest blink of an eye in geological or cosmological terms. It's been just twenty centuries since the dawn of Christianity, and the entire recorded history of humanity only goes back another thirty centuries or so, to the ancient Sumerians with their tablets of clay and cuneiform script.

A hundred years is a long time for a company too. Businesses wax and wane over the years, and while some outfits have been around for several centuries, it's unusual for a company to last that long, even rarer to mark ten decades in the same sector. As industries have risen and fallen, it takes a special kind of strength, vision, agility, and flexibility to adapt from the early days of the industrial revolution, through the rise of heavy industry and steam power, war, the shift to electrical power, and into the digital revolution and internet-led economy of today. Making those transitions from hammering metal and burning coal to managing global production chains at the speed of light over fiber-optic cables is the reserve of truly impressive firms—and the people behind them.

The path isn't always a direct one though. Amazon, one of today's giant corporate names, launched from a garage as an online mail-order bookstore but is now the world's biggest retailer, with an entertainment division and a Web Services operation that provides cloud computing and the backbone structure for a third of the entire internet. Oh, and its founder, Jeff Bezos, is also bidding to provide NASA with the next generation of space rockets. Apple began its corporate life selling home-built desktop microcomputers (complete with DIY

OPPOSITE: This beautiful impressionist image of an R37 by Norman Cucuel was a magazine advert for BMW's bikes and engines from 1925.

ABOVE: At the start of the twentieth century, the aircraft industry was at the forefront of engine design. The IIIa was the first successful powerplant from the Bavarian Motor Works, a 200bhp inline-six with superb high-altitude performance, produced toward the end of World War I.

BMW's most recent Boxer is the R18 mega-cruiser. Seen here in the 2022 Transcontinental tourer and "B" bagger versions, it combines authentic BMW heritage—a giant air-cooled flat-twin engine—with cutting-edge technology and equipment, including a 280-watt Marshall sound system. What would R32 designer Max Friz have made of this, we wonder?

wooden casings), but it's now one of the most valuable firms of all time, selling slick smartphones as part of a futuristic online infrastructure that incorporates entertainment, fitness, work, and home life.

In the motorcycling world, some of the biggest names have taken rather tortuous routes to where they are now. Japan's Yamaha began as an organ and piano manufacturer in the late 1800s (it still makes high-end musical instruments today). Suzuki started as a manufacturer of weaving looms in 1909, while Kawasaki's first enterprise was a Tokyo shipyard in 1878. It could not have been at all clear to the founders that their firms' futures would lie, at least partly, in motorcycle production.

What of BMW? The German firm has followed a slightly straighter pathway over the past one hundred years. Its roots lie in the cauldron of the first World War, when it launched as an engineering firm—the Bavarian Motor Works—in the world of pioneer aviation. It was founded to produce engines in Bavaria, Germany, powering the early warplanes that emerged between 1914 and 1918. Massive inline-six-cylinder aircraft engines like the BMW IIIa were a long way from the small, low-powered, unreliable powerplants used in the pioneering motorcycles of the time. But the design and engineering principles were the same. When the war ended, and Germany was struggling under the terms of the armistice, BMW kept going, finding new products it could make using those design and engineering principles—cars and motorbikes.

The first true BMW motorcycle—the R32—appeared in 1923, half a decade after World War I's end. It marked the beginning of a long journey for BMW Motorrad, a story still developing a full century later. From horrendous low points during World War II, through the hard times of German partition and postwar rebuilding, to much happier recent days, the BMW Motorrad saga has never been dull. It's the story of a company that is welded to its principles and traditions, so much so that its most recent all-new model, the R18, uses the same basic layout as that original R32: an air-cooled flat-twin engine, shaft drive, and steel-tube frame. BMW's unshakable belief in certain design parameters—eschewing a final drive chain, for example—led it down some design cul-de-sacs, like the laid-down inline-three and -four-cylinder K-series machinery of the 1980s. The advantages of that layout in terms of a simple shaft final drive, without a power-sapping, inefficient bevel drive to turn the transmission through 90 degrees, was admirable in its dedication. But the compromises it generated elsewhere in the machine design—particularly in the chassis—became unjustifiable in the late 1990s.

In other areas, though, the firm's commitment to certain technologies and design principles paid off handsomely. Fitting fuel injection and then ABS to a motorcycle like the K100 in the 1980s seemed like a pointless exercise in overengineering, complexity, and cost. The Munich firm ploughed a lonely furrow with both of those foundation technologies for

ABOVE: **The laid-down four-cylinder engine in BMW's K100 was a unique choice and made for a very different motorcycle from mainstream transverse layouts. But it was an advanced design, with DOHC head and Bosch fuel injection from the very start.**

BELOW: **The author testing the most recent BMW Boxer, the R18 Classic, at a press launch in 2021. Leather saddlebags and windshield add touring practicality with a dash of retro style.**

BMW's obsession with innovation meant it was the first to sell an ABS-equipped motorcycle in 1988, on selected K100 models. On this full-touring LT version, the ABS pump is prominently mounted behind the rider's leg on the right-hand side.

another decade or so, before they became essential to comply with emissions and safety laws across all the big markets. Meanwhile, the most successful of the firm's bikes—the R1100 to R1250 GS range of big adventure machines—stemmed from an incredibly unlikely off-road design, the original R80 G/S. While the competition was building small, lightweight, narrow machines with chain drive and powerful single-cylinder engines, BMW offered a heavy, solid, wide flat-twin, with a massively heavy shaft final drive. And rather like the tortoise and the hare, those mainstream dirt bikes fell by the wayside as the GS defined an all-new range of top-selling bikes and the other firms rushed to catch up again.

Even more impressive, perhaps, is the way BMW as a company has maintained its independence. Virtually all its peers in the automotive world have undergone takeovers, changing administrations, liquidations, mergers, and buyouts over the last 125 years or so of car production. The Bavarian Motor Works has largely escaped those fates despite a number of tough periods in its history, and it remains a publicly traded company. The BMW AG parent group owns the Motorrad business alongside its BMW car manufacturing division, as well as the Rolls-Royce and Mini brands.

Arguably the most exciting part of the BMW story has been the last twenty-five years. I've worked as a motorcycle journalist throughout this period and watched the bikes from the firm's Berlin factory get better and better every year. When I started riding in the late 1980s, the BMW badge held little appeal for me, an eighteen-year-old biker in Glasgow, Scotland, obsessed with Kawasaki GPZs, Yamaha FZRs, Honda VFRs, and most of all, Suzuki GSX-Rs. Those sexy high-powered sport bikes ranked alongside supercars like the Toyota Supra, Canon's cameras, and Sony's audio systems as symbols of the might of Japanese design and engineering. Sure, German engineering was impressive in its own way; Mercedes-Benz, Leica, and Grundig enjoyed immense cachet. But when it came to bikes, BMW, as the only German option, looked a bit old and dull in comparison to Japan or Italy. The firm that built 1,500-bhp turbocharged Formula One engines and the mighty M cars had nothing to match the Kawasaki ZZ-R1100, the Honda VFR750R RC30, or even Yamaha's TZR250 in terms of pure sporty performance on two wheels.

By the time I began work on this book, however, that was all ancient history. Beginning with the K1200 S in 2004 and continuing with the S1000 RR in 2009, BMW reinvented itself as a high-tech, high-performance bike builder that could match—

ABOVE: **The ancestor of the modern GS range is the 1980 R80 G/S, an ungainly variant of the R80 road bike with long-travel suspension and dual-purpose tires.**

LEFT: **The 2004 R1200 GS was a long way from the R80 G/S, but it stuck with the same basic layout: boxer engine with single-sided shaft drive swingarm, enduro suspension, and dirt-friendly tire options.**

and top—anything from Japan or Italy. Pioneering technologies like electronic suspension control, ride-by-wire engine management, gear quickshifters, and novel suspension designs pushed the bike world forward, while the sheer usability and effectiveness of the GS range made them into a global success story.

At the end of its first century, BMW is in a fantastic place. In some ways, though, the challenges before it are almost as daunting as they were in 1923 (though very different). The internal combustion engine is on its way out, with governments around the world stipulating an end to the use of fossil fuels. Hydrogen fuel cells, new battery technology, and electric motors are already appearing in BMW cars—but bikes are a trickier proposition. With none of the space or mass margins available to automobiles and trucks, making a 200kg, 150-bhp bike that can travel 150 miles between "fuel" stops and refill or charge in minutes rather than hours is currently impossible. And with self-driving cars, working remotely, and other societal changes altering how people move, the purely practical side of motorcycling for commuting and urban transport is also evolving. But, if any firm can manage it, you'd have to put your money on the smart folks at the Bavarian Motor Works as they head into their second century.

ABOVE: **This was the first sign that BMW was becoming serious about performance again. The K1200 S engine produced extreme levels of horsepower, and the chassis was just as innovative, with new Duolever front suspension, electronic suspension adjustment, and an electronic gear quickshifter.**

RIGHT: **With the S1000 RR, BMW benchmarked the best that Japan could offer in the 1,000cc superbike class, then matched it with its own machine. It was a more conventional design than typical BMW solutions, but radical looks and high-tech equipment marked it as something special.**

OPPOSITE: **The 2022 CE 04 urban maxi-scooter is BMW's first attempt at a mainstream production electric bike, combining 400cc-class performance with futuristic styling. BMW also designed a matching "smart" jacket that has an integrated Bluetooth touch control panel for the Motorrad phone app and LED safety lighting built in.**

RAPP
MOTORENWERKE
G.M.
MÜNCHEN.

Origins

There's an internet meme about a *London Times* newspaper story in 1894, predicting that within a few decades, the streets of London would be nine feet deep in horse manure. It was estimated that fifty thousand horses worked in the British capital at the turn of the century, leaving hundreds of tons of manure, urine, bedding, and feed spread throughout the city. The resulting filth, flies, maggots, and dead horses threatened to block the pace of urban progress, and it was the same in New York, Paris, Munich, and Turin.

Luckily for the early Edwardian city folk of London, technology was about to come to their rescue in the form of the internal combustion engine (ICE). It was a perfect confluence of scientific and engineering advances: experimental ICE designs had been seen in laboratories from the late eighteenth century onward. Then, in the 1860s and 1870s, Nicolaus Otto transformed the ICE from a lab curiosity into an industrial game-changer, similar to how Scottish engineer James Watt had converted the steam engine a century before. Otto produced a series of ever-improving designs, culminating in 1876 with a four-stroke, spark-ignition, compressed-charge design—the Otto cycle engine—which is still used in virtually every car and motorcycle engine on the road today.

Otto's technology was borrowed and adapted by one of his business partners, Gottlieb Daimler, and used in the 1885 Daimler Reitwagen—arguably the first "motorcycle." It had four wheels (two were just stabilizers), and while the chassis had a ways to go, it was the first single-track vehicle to use a gasoline-powered Otto-cycle engine. The world was at the start of something great.

The major focus of the Otto cycle engine lay elsewhere. Fellow German engineer Karl Benz developed his own engine and put it into the first automobile—the Patent Motorwagen—in late 1885/early 1886. By the turn of the century, new cars, buses, and trucks powered by the internal combustion engine suddenly appeared. Rudolf Diesel's development of his eponymous compression ignition engine, also in Germany in 1892, broadened the appeal of these new powerplants, while accompanying developments in crude oil refining and metallurgy provided better fuels and stronger, tougher metal alloys.

Karl Rapp's engineering firm supplied engines to the Otto airplane company and was another main pillar of the BMW firm that emerged at the end of World War I.

ABOVE: Gustav Otto's father, Nicolaus, is the man behind the four-stroke gasoline engine that transformed the twentieth century. Gustav didn't have quite the same impact, but his Otto Flugmaschinenfabrik aircraft firm is one of the foundations of the BMW brand.

RIGHT: If this were a book about BMW cars, we'd be waxing lyrical about this being an essential part of the firm's powerplant DNA. The IIIa aircraft engine was a straight-six design, the layout that powered so many excellent BMW automobiles. The layout is limited as a useful motorcycle engine, but BMW has used a straight-six motor in the K1600 super-tourer range since 2010.

Germany, the Powerhouse behind the New Powerplant

Germany was right at the heart of these early developments. There was a group of engineers, scientists, and businessmen—including Benz, Otto, Daimler, and Diesel as well as Wilhelm Maybach, Robert Bosch, and (a little later) Ferdinand Porsche—all pushing forward the evolution of the internal combustion engine.

Germany would also, indirectly, give ICE technology another massive push as part of the unimaginable tragedy that was World War I. At the beginning of the Great War in 1914, transportation outside the steam railways was almost entirely muscle-powered: horses and men schlepped big guns, bullets, and bully beef to the front line. But by 1918, just four years later, gasoline was powering tanks and trucks—and airplanes.

Heavier-than-air flight had begun only a decade earlier in December 1903, when the Wright brothers flew their Wright Flyer at Kitty Hawk in North Carolina. The Flyer had a home-brewed inline-four engine with just 12 bhp, but as airplane technology took off, the demand for more powerful designs accelerated. Engines grew in capacity, and the number of cylinders increased to power the

pioneering fighters, bombers, and reconnaissance planes doing battle above the Western Front.

Halfway through that terrible war, Bavaria in southern Germany emerged as a center of excellence in terms of aircraft and engine design. The Rapp Motorenwerke firm had been founded in Munich by engineer Karl Rapp in 1913 to build inline-four aircraft engines for the Otto Flugmaschinenfabrik aircraft builder set up by Gustav Otto, son of

The Boxer Engine

BMW's first Boxer engine wasn't an aircraft design, despite the firm's roots in that industry. Ironically, however, the flat-twin layout was an excellent arrangement for piston-powered aircraft engines. Having the cylinders sticking out from each side allowed for optimal air cooling, saving the weight and complexity of a liquid-cooled inline engine. It had a low profile, so the pilot could easily see forward over the top of the powerplant and through the propeller arc. And the crank was perfectly in line with the propeller shaft, simplifying power transmission.

Most of these benefits are also useful on a motorcycle. Air cooling was standard in the early years of bike development, avoiding the cost, complexity, and reliability problems associated with early liquid-cooling technology. And layouts that lent themselves naturally to air cooling had an advantage. (The Boxer layout even kept riders' feet warm, a welcome benefit.)

There were other advantages to the Boxer. It kept much of the weight of a bike down low for a lower center of gravity. That reduced agility but improved stability and made the weight of a large bike easier to handle. Having the crankshaft oriented fore and aft allowed a shaft final drive setup without expensive, inefficient drive transfers through 90 degrees, as with a transverse engine. Back in the early 1900s, drive chains were weak and unreliable compared with the high-tech O-ring-sealed designs of today, so a simple steel drive shaft was far more reliable and efficient, though pricier to engineer initially.

Finally, the flat-twin Boxer had good balance. The pistons moved in opposite directions, canceling out much of the vibration caused by their motion. (That movement of the pistons is where the Boxer name comes from: the in-and-out motion emulates the action of a revved-up fighter in the boxing ring, punching their fists together in front of their chest.)

BMW's first bike, the R32, used a 494cc Boxer engine with a basic side-valve layout. The long-stemmed valves were mounted parallel to the cylinders, facing outward, and operated by a central camshaft. The inlet and exhaust ports were off-center, with the spark plug and combustion chamber positioned above the cylinder bore. This was fine for the day, as low-octane fuel limited compression ratios, but it's terrible from a modern point of view in terms of combustion chamber shape and efficient gas flow.

Over the one hundred years since, the Boxer has been developed almost beyond recognition. The latest R1250 Boxer engine has DOHC four-valve heads, variable valve timing, water cooling, and ride-by-wire fuel injection, and it makes fifteen times the power of the R32 engine from two and a half times the cubic capacity. But it still has cylinders sticking out of each side and shaft final drive—and it still keeps your feet warm on cold days.

ABOVE: The R32 motor looks utterly archaic, closer to the steam engines of the nineteenth century than the advanced powerplants of today. Note the two bronze plugs where the side valves are located, the hand-gear change, and the inlet and exhaust ports set off from the combustion chamber. The advantages of the shaft drive to a Boxer engine are clear: it's in line with the crankshaft, giving a direct, simple route transferring power from engine to rear wheel.

BELOW: This cutaway model of the latest R1250 Boxer engine shows how much more complex the design has become. The four-valve DOHC heads, the ShiftCam variable valve timing, the compact clutch, water pump and vertical fuel injection stacks—it's all a long way from the one-hundred-year-old R32. But at its heart, the latest BMW engine is still a shaft-driven Boxer twin.

BMW
25001
B M W

The M2B15 was the first non-aircraft engine produced by BMW, but it wasn't fitted to a BMW bike. Rather, it was supplied to the Victoria motorcycle firm and was also used in the Helios bike. It formed the basis of the M2B33 used in the R32, the first bike to wear the BMW logo.

Nicolaus. Following reorganization, Otto's company was renamed Bayerische Flugzeugwerke (Bavarian Aircraft Works) in 1916, while Rapp left his business and its name changed to Bayerische Motoren Werke—the first use of the BMW brand—in 1917.

The first ever BMW engine appeared that year—the IIIa aircraft engine. It was advanced for the time: a water-cooled SOHC inline-six with a 19-liter capacity and 180-bhp output. It powered Germany's most advanced Fokker and Junkers warplanes and was generally recognized as the best of its kind. The success of BMW's IIIa engine ensured a rapid expansion of the company at its base in Munich, with a new factory built near the Oberwiesenfeld airfield, close to Otto Flugmaschinenfabrik. However, the end of the war and the Treaty of Versailles put an end to that nascent aircraft industry. The terms imposed by the victorious Allies banned Germany from making any airplanes at all for six months after the Treaty was signed, until the middle of 1920.

Planes No More

BMW had to find something else to do. Max Friz, the firm's design director, and Franz Josef Popp, general director, dabbled with producing truck engines based on the IIIa aircraft engine layout. They also supplied a 6.5-bhp, 494cc air-cooled flat-twin engine, called the M2B15, to the Victoria Werke motorcycle company in Nuremberg. The M2B15 had been designed by Friz alongside Martin Stolle, a senior BMW engineer who also loved motorcycles. Stolle owned a British Douglas Model B motorbike, with a 544cc flat-twin engine, and this provided the inspiration for the new 6.5-bhp Boxer engine.

The M2B15 also powered a bike called the Helios, built by the Bayerische Flugzeugwerke company. BFW would later merge with BMW, so the Helios is definitely part of the firm's ancestry—though it was reportedly an appalling piece of design. Friz famously wanted to throw it in a lake,

The 1919 Helios was built by Bayerische Flugzeugwerke, BMW's predecessor firm. So, while it wasn't a "true" BMW, it certainly deserves a part in the firm's history. Note the rather illogical installation, with the cylinders arranged fore and aft, rather than out in the cooling breeze. That's a consequence of the chain final drive, different from the shaft used by BMW.

1923–1926 R32

The very earliest motorcycles look utterly alien to a modern motorcyclist's eyes. The 1894 Hildebrand & Wolfmüller, the first mass-produced motorbike, didn't even have a crankshaft— the piston con rod was connected directly to a crank on the rear wheel, like a steam locomotive. Even twenty years later, machines like the 1915 Triumph Model H looked more like bicycles with a crude engine attached.

But the BMW R32 was different. From a distance, it could pass for the sort of modern custom BMW you might see parked outside the Bike Shed in London or beautifully photographed on the BikeEXIF website. It had a Boxer engine that looked part of the overall design. There's shaft final drive and a fairly robust-looking steel-tube cradle frame. Granted, the suspension and brakes looked rather odd, and the hand gear change was a bit of a worry, but that brown leather seat was all the rage in London and Milan that season.

It's a testament to the engineering and design skills of Max Friz and his team that this first attempt was so close to modern motorcycling sensibilities. While it had no rear suspension or front brake, the 8.5-bhp engine gave a top speed of nearly 60 mph, which was more than what most roads at the time could handle. Best of all was the design and quality of the components. The guys who learned their trade building aircraft engines knew all about making things properly, and the result was a premium, reliable, effective piece of kit that was far ahead of the competition at the time.

The engine had one-piece cast-iron cylinders with an integrated head, screw-in bronze covers over the side valves, aluminum pistons, and gear-driven camshaft. The engine lubrication used a wet-sump gear pump system rather than the total-loss setups common elsewhere, and the ignition was via a Bosch magneto mounted above the front of the crankcases.

That cutting-edge powerplant was bolted into a steel-tube cradle-type frame, with a solid back end that simplified the shaft final drive. Front suspension was via a trailing link fork with linkage to a short leaf spring mounted below the steering stem. There's no front brake on the original R32, just a heel-operated block brake for the rear wheel, but later variants got a front drum. Electric lighting was an option, and there was a complex-looking belt-operated cable drive to the speedometer.

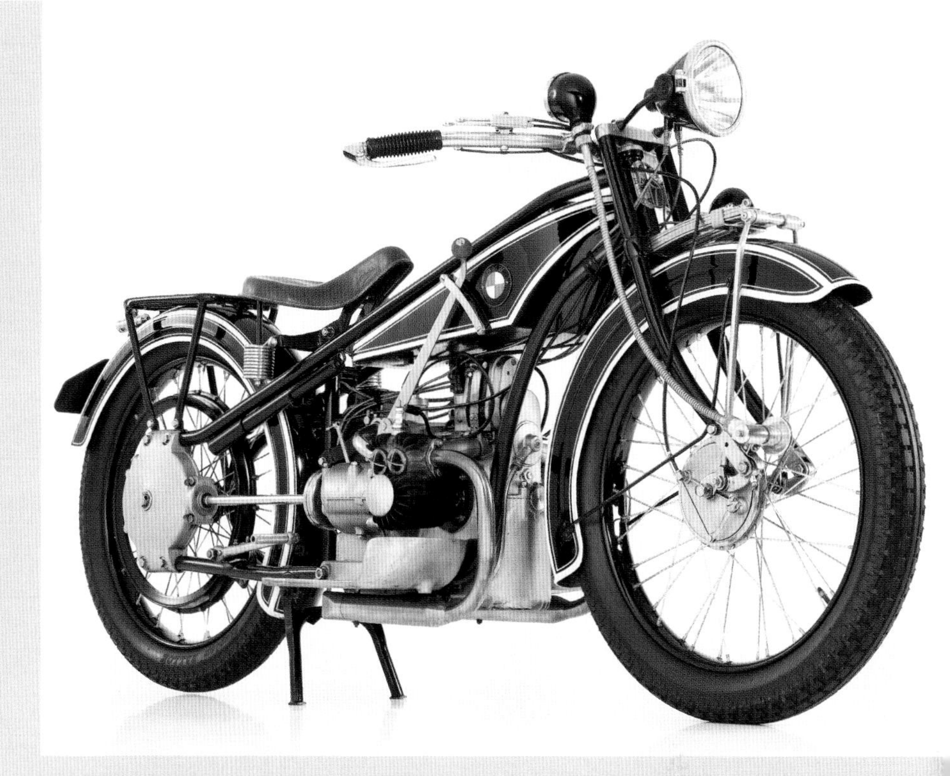

This stunning BMW photo lets you have a really close look at what a motorcycle was like in the early 1920s. There are some standout oddities: the hand gear change, the solid rear end, and the leaf-spring front suspension. But the Boxer engine, steel-tube frame, and shaft drive all mark the R32 as the archetypical BMW, the ancestor of every R-series Boxer since.

A sprung seat added comfort, and the combination of a sealed, oil-tight engine and shaft final drive, plus extended footboards with a front toe shield, meant the R32 rider about town would arrive much cleaner than on other, more primitive machinery of the time. It was a premium bike, costing 2,200 Reichsmarks, and the firm sold more than three thousand of them during the production run.

The R37 came later, a high-performance development of the R32, with a new overhead-valve version of the M2B33 engine. The tuned engine was called the M2B36 and made a healthy 16 bhp—nearly double the power. Aluminum alloy OHV cylinder heads gave the engine a very modern look, with the enclosed valve covers keeping the top end clean and lubricated. Aimed at competition and high-performance use, it was initially used at the famous Solitude races near Stuttgart in 1924 and won the German championship under Franz Bieber that year. It was then released as a production model in 1925. It could hit 70 mph but was pricey at 2,900 Reichsmarks, so only 152 units sold.

One of the engineers behind the M2B15, Martin Stolle, is shown here on the Victoria bike that used his engine. Again, like the Helios, the engine is mounted the "wrong" way, this time to accommodate a belt final drive.

1925–1927 R39

The R32 was a great bike, but its advanced, well-engineered design priced it beyond many customers. BMW needed a cheaper entry-level option, and the R39 was the answer. The firm removed one cylinder and rotated the crankcase 90 degrees so the cylinder pointed upward, then fitted it with the latest OHV alloy cylinder-head technology.

It's important to note that the crank on this bike was still fore and aft as on the Boxer design; it didn't have a conventional single-cylinder design with a transverse crank. That allowed the same transmission and shaft-final-drive layout as the twins and simplified the production process. The R39 made less power (6.5 bhp) than the R32 but was smaller and lighter, so performance was about the same. It was also cheaper to buy—500 Reichsmarks less than the R32 at 1,870RM—but wasn't a great success. It was discontinued after just two years.

The R39 was BMW's first attempt at a cheaper entry-level single-cylinder machine. It used half a Boxer engine, rotated through 90 degrees, and stuck with shaft drive and fore-and-aft crankshaft. It wasn't a big hit, but BMW kept building singles like this until the late 1960s.

1926–1928 R42

Just like modern bikes, the R32 needed a bit of an upgrade three years after it launched. The R42 was the answer: it stayed with a 494cc side-valve Boxer lump, but a redesigned top end with separate alloy heads and other mods meant it made 50 percent more power, a heady 12 bhp at 3,400 rpm. The chassis was overall similar to the R32 but with more brakes (a good thing) plus frame and suspension tweaks. The best mod was to the price: it was 30 percent cheaper than the R32 and sold like hot *apfelstrudel* to enthusiastic late-1920s riders. More than 6,500 of them left the Munich factory in two years, double the R32 sales figures.

The 1926 R42 was the first major overhaul for the R32. It gained alloy cylinder heads, an extra 4 bhp, and a cheaper price tag.

BMW's first race replica was the R37, an OHV version of the R32 with alloy cylinder heads, pictured here with legendary BMW engineer Rudolf Schleicher.

but his boss encouraged him to go one better and redesign it, which he did within a month.

The next few years were difficult for everyone in Germany. But as the economy recovered somewhat and society stabilized, BMW was able to focus efforts on its medium-term future. Aircraft engines were still out of the question, so the firm survived by making agricultural equipment, truck and boat engines, and even brake components for trains. Behind the scenes, the BMW and BFW businesses were the subject of various takeovers, financial rearrangements, and other byzantine ownership changes. The end result was a single firm, called BMW, with a factory on the outskirts of Munich and plans to build small cars—and motorcycles.

The M2B15 was coming to the end of the line. Its main customer, Victoria Werke, had switched to an alternative OHV engine supplied by the Wilhelm Sedlbauer firm, which had taken on Martin Stolle after he left BMW. Friz and Popp needed something new, and the result was the revolutionary R32, launched at the Paris show in late 1923. It had a proper steel-tube cradle frame, leaf-spring front link suspension, a single band-type rear brake, and an upgraded version of the M2B15 motor, the M2B33. The engine was mounted across the frame for better cooling, with a shaft final drive to the fixed rear wheel. It made 8.5 bhp, weighed 122kg, and was a worthy beginning to the BMW Motorrad story.

Horsecrap to BS

That internet meme about the *London Times* horse manure story is, like a lot of stuff on the internet, bullshit rather than horseshit. The *Times* has no record of such a story in its archives, but the point it illustrates remains the same. Predicting how the future will look is a fool's errand in the best of times, but if you are going to try, assuming linear change is a big mistake. Disruptive technologies like the internal combustion engine can completely change how the world looks. A century on, our cities are indeed clogged up and suffer from terrible pollution. But rather than nine feet of horse manure, it's produced by cars and trucks using gas and diesel engines.

1927–1930 R47/R57

The R47 was the upgrade on the R37 "race replica," with even more power from the 494cc OHV motor (now making 18 bhp) and a 1,000RM price cut. The chassis gained a new rear brake design that worked directly on the drive shaft rather than the rear wheel, and it weighed 130kg dry. The lower price tag meant the firm sold more than 1,700 of them, and it was a common sight at early motorcycle race meets across Europe.

Another overhaul for 1928 produced the R57. Price, power, and engine design remained largely the same as the R47, but with a slightly larger 24mm carburetor and a 200mm front drum brake (a 50mm increase on the R47).

The R47 appeared in 1927 as an update for the 1937 race-rep. More power and a lower price tag made it the privateer racers' choice across Europe in the late 1920s. ***Henry von Wartenberg***

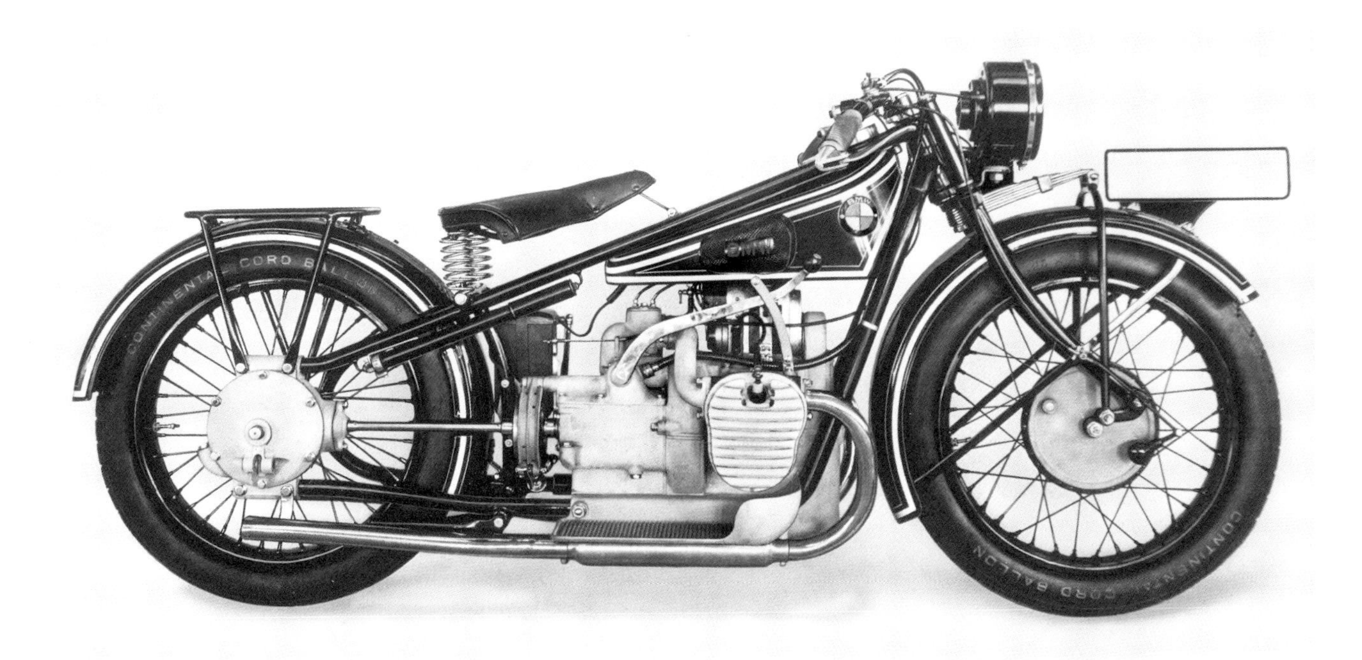

1928–1929 R52

On the face of it, the R52 looks like a logical update to the R42 (itself the R32 upgrade). But the R52 used an all-new 486cc engine design, with a long-stroke 63x78mm architecture. Power remained about the same at 12 bhp, as did the top speed of around 62 mph, but a higher dry weight of 152kg reduced overall performance. The front brake was the new, larger 200mm drum used across the range, and it was supplied ready to use as part of a sidecar outfit, which was increasingly common in the late 1920s.

To modern eyes, it looks like another mad old black BMW classic, but the R52 was a heftier update, with a new long-stroke side-valve engine and a proper drum front brake of all things. It was a great match for the sidecars, which were becoming more fashionable.

Roaring Thirties

By the end of the 1920s, BMW's position as a successful engineering and manufacturing company had been massively strengthened. The period of economic tumult after the Great War, together with the successful Bolshevik revolution in Russia (plus a failed revolution in Germany itself), had begun to recede, and the economic madness of Weimar Republic hyperinflation was gone. A new currency (the Reichsmark) and relatively stable government meant that the second half of the 1920s was a calmer, more prosperous time. But clouds were gathering.

The reparations imposed on Germany by the Versailles Treaty were a terrible economic burden, and the political arena was becoming more polarized between left and right— the Communist Party of Germany (KPD) and Adolf Hitler's National Socialist German Workers' Party (NSDAP), the Nazis.

In Munich, BMW resumed airplane engine production after the Versailles restrictions on military and aircraft production were lifted. The original IIIa straight-six aero-engine returned to production in 1923, and 1926 saw the launch of a monster 47-liter, 750-bhp V-12 engine, the BMW VI. By 1929, BMW had gained a license to build American Pratt & Whitney radial engines, and Germany's flag-carrier airline, Lufthansa, was insisting on BMW-built engines for its planes.

The firm also moved into car production, taking over the Fahrzeugwerke Eisenach automaker in 1928 and producing a license-built version of the British Austin Seven, called the Dixi 3/15. The car factory in Eisenach, central Germany, would later take on R75 motorcycle and sidecar production in the mid 1940s when the Munich operation switched fully to wartime aircraft engine production.

One of the great names of BMW's early bike engineering, Rudolf Schleicher, shown here re-united with an R37, back in 1987.

Bikes Remain the Mainstay

Even with major advances on the airplane and car fronts, bikes remained at the heart of the business. The late 1920s and early 1930s saw a slew of new and updated models, showcasing new tech in engine and chassis design as well as engineering. The new long-stroke 500 motor of the R52, plus the high-performance OHV sporty 500 used in the R37/47/57, gave BMW a fairly broad range of high-quality Boxer twin engines, as well as the entry-level R39 250 single. Meanwhile, on the chassis front, a larger drum front brake and a rear brake that acted upon the drive shaft rather than the back wheel provided much more

ABOVE: **License-built radial engines like the ones on this Junkers Ju52, alongside the firm's 750 bhp, 47-liter V-12 VI engine, helped BMW soar to the top of the aircraft engine marker in pre-WWII Germany.**

RIGHT: **BMW's first successful car design was a licensed version of the British Austin Seven, built at a new factory in Eisenach, central Germany. Eisenach would later switch to bike production during World War II before being taken over by the Soviet Union when Germany was partitioned into East and West.**

1928–1929 R62/63

BMW itself reckons that the R62 and R63 marked a new era for the firm. The upgrade to a 50 percent larger 750cc engine was a big move, but the firm also developed two versions of this big-bore powerplant. The R62 used the M56 variant, an older, more conservative 745cc side-valve design with a square bore and stroke, making 18 bhp. The R63 got the M60 engine, a racier, short-stroke OHV 735cc layout that produced an astonishing 24-bhp. Both bikes used similar steel-tube cradle frames, and the R62 was often paired with a large sidecar, with lower final drive gearing available to suit. The three-speed hand-operated gearbox remained, and the transmission now used a proper oil lubrication system instead of just packed grease.

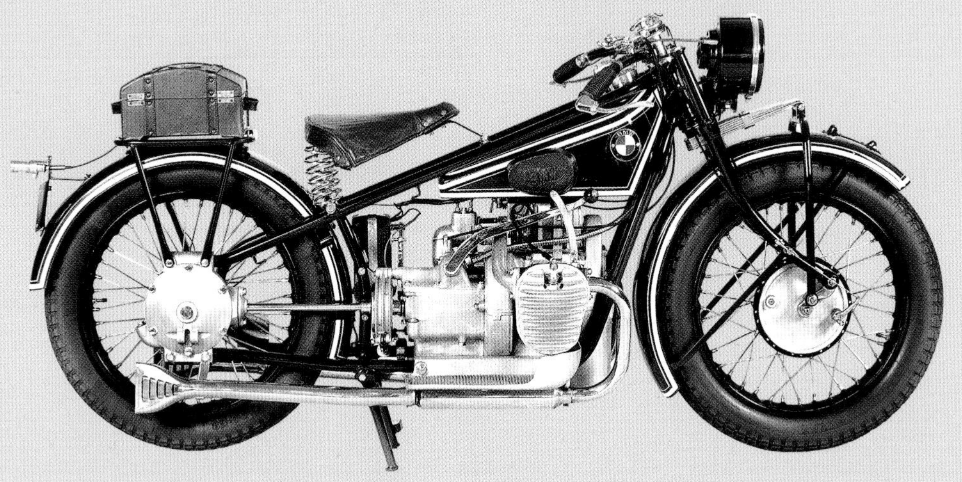

The R62 and R63 formed a small lineup of 750cc machinery, but they had very different engines. The R62 shown here had the more basic side-valve engine that had better low-down drive, while the R63 had a OHV short-stroke powerplant with stronger top end. The R62 was ideal for sidecar use, while the 24-bhp R63 offered a more sporting ride. ***Henry von Wartenberg***

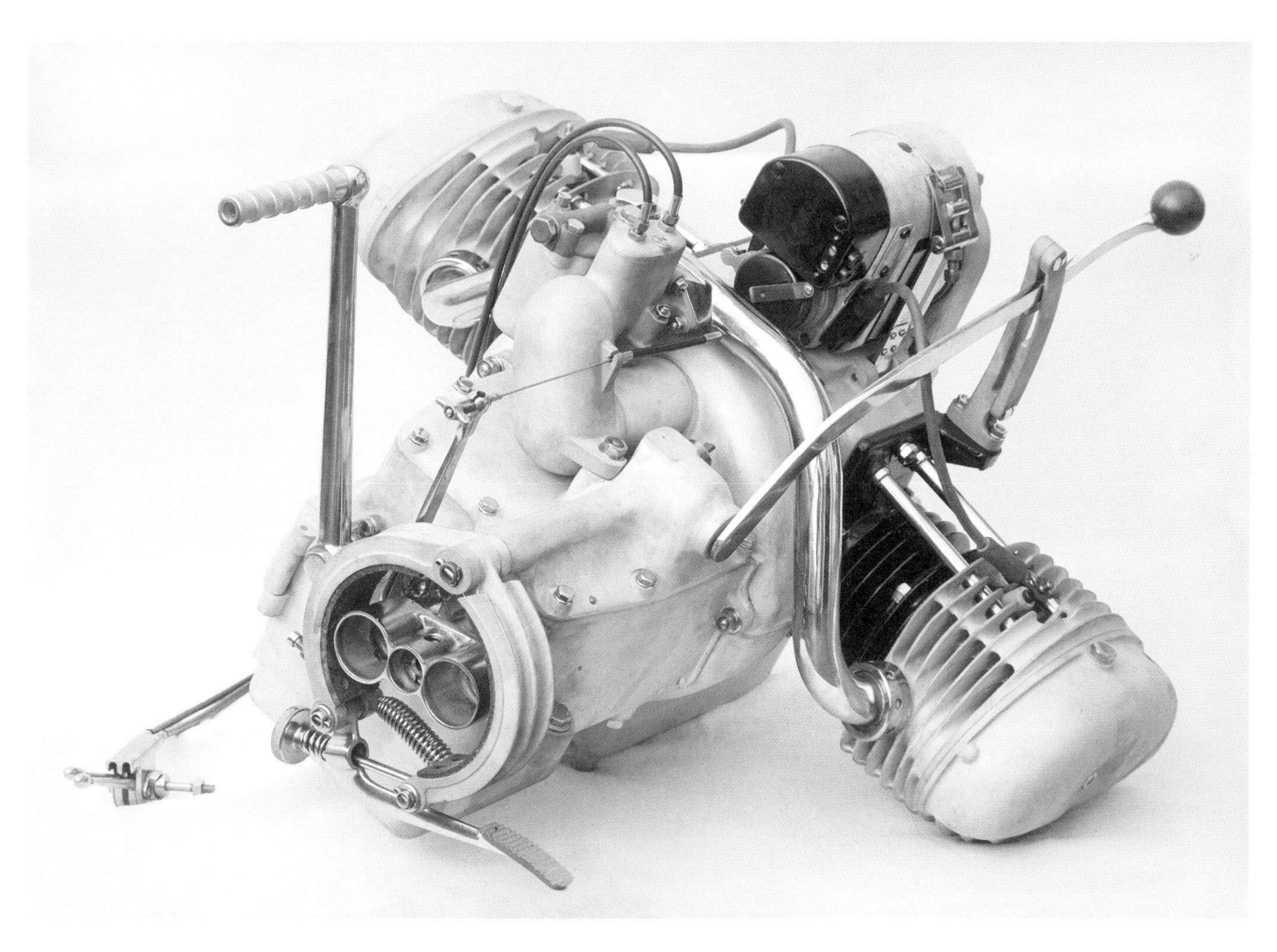

This unusual angle shows the R63's shaft-drive brake shoes, as well as the outward-swinging kick start BMW used on all its early models.

Side Valve versus OHV

It's a little difficult for a modern mind to appreciate the constraints on engine designers in the nascent days of motorcycle and car design. Back in the early 1900s, there were no CNC machine tools, no advanced aluminum casting methods, and no five-axis mills in every workshop. These pioneering engineers had to use simpler, basic engine designs that were practical for mass production with the tools and skills available.

Hence the side-valve engine layout used on the earliest BMW Boxer engines. From a twenty-first-century viewpoint, it looks insane: the valves are "upside-down" and sit parallel to the cylinder bore, with long stems reaching down to a camshaft centrally mounted above the crankshaft. The intake and exhaust ports sit above the cylinder axis, leading into a large offset combustion chamber. It works—and avoids the cost of a more complex valve train. There are several downsides, however: it's difficult to get a good combustion chamber shape (the ideal chamber would be spherical with a spark plug in the center), and compression is low because of the large offset ports. Low-compression engines are less thermally efficient and produce lower levels of torque and power, while the slower combustion across a large, unevenly shaped chamber limits maximum revs.

A much more efficient solution is the OHV design used on the earliest high-performance Boxers. Here, the side-mounted valve stems are replaced by long pushrods that lead to valve rockers, small levers that take the pushrod movement and reverse it, pushing down to open spring-loaded poppet valves mounted above the head. The intake and exhaust ports are in line with the cylinder bore, and the combustion chamber is directly above the piston crown, allowing a much more compact chamber, efficient fuel burn, and higher compression ratios. The result is more power and torque, and higher efficiency, at the cost of a more complex cylinder head design.

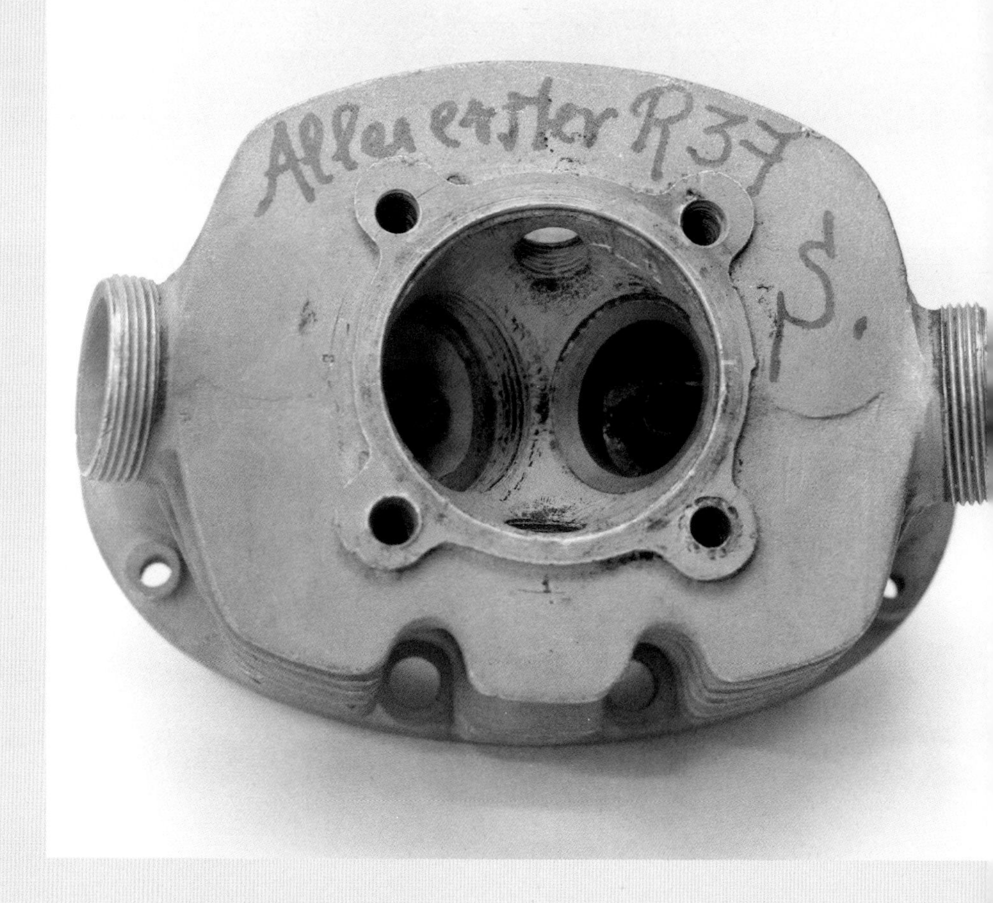

This R52 engine has the cylinder head removed, showing how a side-valve layout is arranged. The valves open outward into a large asymmetric combustion chamber; inlet and exhaust ports run above the cylinder. The R37 head is an OHV design, with the valves opening inward toward the top of the piston, which allows a much more compact combustion chamber.

Franz Josef Popp, seen at right, had been part of BMW's senior management from the very start and helped design the R32. During the 1930s, he had to tread a fine path between the interests of the Nazi state and those of the BMW company. Fail to keep the Führer happy, and the entire firm could be nationalized with the stroke of a pen.

stopping power. Annual sales had more than tripled in just five years, from 1,640 bikes sold in 1924 to nearly 5,700 in 1929.

It wasn't enough for chief engineer and design director Max Friz though, and for 1928, a larger 750cc class emerged in the R62 and R63 heavyweight machines. The R62 was the tourer, and the R63 was sportier, and they used two completely different powerplants to suit.

Depression, Then the End of Democracy in Germany

Through the early 1930s, the world experienced the biggest economic crash of the twentieth century. From Wall Street to the City of London and the Munich Börse, banks failed, businesses crashed, stock markets plummeted, and unemployment ravaged Europe and the United States. Germany didn't escape any of this, and the economic disaster spilled over into the political arena with disastrous results.

The democratic Weimar Republic era ended in January 1933, when President Paul von Hindenburg appointed Adolf Hitler as chancellor. Hitler's NSDAP Nazi party had gained the largest number of seats in the Reichstag elections of November 1932 thanks to growing support for his populist policies among unemployed youth, the lower-middle class, and rural voters. Hitler and the Nazi party rapidly suspended democracy and the rule of law, imposing a totalitarian regime, and began preparing for war.

Power and Propaganda

In economic terms, BMW as a company benefited from rearmament. Senior managers like Franz Josef Popp, general director of BMW from 1922 to 1942, were expected to become Nazi party members, and there was a risk of the firm being nationalized and taken over completely by the Nazi government. There was little choice but to go along with the business of war preparations: aircraft engines for covert rearmament and essential cars and

BMW

motorcycles for the growing state apparatus of armed forces, police, and Nazi party officials.

The firm also benefited from Hitler's enthusiasm for national achievements in sport and competition, including car and motorcycle racing and land-speed records. With official blessing for assaults on the Isle of Man TT and motorcycle speed records, BMW developed two supercharged machines: the WR750, powered by an OHV Boxer, and the more advanced Type 255 Kompressor 500, a high-tech 492cc Boxer engine with bevel-drive DOHC cylinder heads. The WR had some success under factory rider Ernst Henne in the motorcycle speed record class, but it was the 60-bhp Kompressor 500 that took BMW to another level. It set a speed record of 173.68 mph in 1937 under Henne and won the Isle of Man Senior TT in 1939, ridden by Georg Meier (the first foreign rider to win a TT race).

ABOVE AND OPPOSITE: **Another all-new engine appeared for the 1936 R5, this time a twin-cam Boxer with two camshafts mounted in the crankcases, allowing shorter pushrods and stronger engine performance. The chassis was also all new, and with a heady 24-bhp peak power, the R5 showed the pathway to future high-performance Boxers.**

On the Road

Away from the rarefied atmosphere of international motorbike-design-as-propaganda, BMW's road bikes kept improving despite the economic and social woes in Germany. The R11 and R16 appeared, bringing a new

BMW's first 600cc machine, the R6, was a low-revving sidecar lugger, featuring an old-school side-valve engine layout.

1935–1937 R17

While 33 bhp isn't much power today, in 1935, it was right at the top of the motorcycle power league. The R17 didn't just have fire-breathing power levels from the 736cc OHV Boxer twin though: it came with a revolutionary hydraulically damped telescopic front fork, the first on a production bike. Other firms had used telescopic forks before, but this was the first with oil damping built into the design, and it made a massive difference to the chassis performance. Like the R12, the R17 used the new pressed-steel riveted frame layout, but it still had a rigid hardtail rear end—a surprisingly basic setup for the era, considering the high-tech forks and powerful engine.

For the first time, BMW offered a bike with proper hydraulically damped front suspension. The R17 was the first production bike with hydraulic front forks, though it stuck with a rigid rear end and a heavy pressed-steel frame design.

Supercharging

The easiest way to get more power from a motorcycle engine is to make it bigger. A 1,000cc engine will always make more power than a 500cc engine of the same general design; it can pump more air and fuel into and out of its cylinders, and more fuel burning means more power at the back tire.

To make an engine bigger in the conventional way, you can use bigger pistons and/or a longer stroke on the crankshaft, to physically increase the swept volume inside the cylinders. But you can also "virtually" increase the capacity of an engine by pressurizing the intake air. If you have a compressor that increases the pressure of the air going into the cylinder, you have more oxygen inside the cylinder and can add more fuel to burn, increasing the size of the "bang" and the force applied to the piston.

There are two types of intake charge compressors: superchargers and turbochargers. Both pressurize the air going into an engine. A turbocharger uses the energy from the engine's exhaust gas flow to spin a turbine that drives the compressor, while a supercharger uses a direct drive from the engine.

Both have advantages and disadvantages, but BMW chose a supercharger for its WR750 and Type 255 Kompressor race machines in the 1930s. The firm used a sliding-vane design, which has an oval chamber with an offset rotor and sliding vanes mounted in slots in the rotor. As the engine turns the compressor, the vanes are flung out by centrifugal force to the inside face of the oval chamber, drawing in the intake air before compressing it as the vanes slide back in. Castor oil lubricates the movement of the vanes, and the pressurized air is forced into the Boxer engines' cylinders, giving a big boost to torque and power all the way through the rev range.

This extra power gave the BMW racebikes an advantage in peak power over their (mostly) British competitors and canceled out the traditional handling advantage held by the Nortons and Velocettes of the time. That (plus some incredible skill) helped Georg Meier to his Senior TT win in 1939. Postwar, forced induction was banned in almost all motorcycle race classes, and BMW wouldn't see another solo TT win until the twenty-first century.

This picture shows the BMW 255 Kompressor's disassembled supercharging unit. The eight "razor-blade" plates are the sliding vanes, which engage in the slotted rotor bottom right, inside the rotor housing. Note the offset mounting points for the rotor shaft.

It sounds like an apocryphal tale, but BMW's own history backs it up: Hitler himself ribbed BMW staff about the lack of rear suspension on its bikes. Competitor machines had already adopted sprung and damped back ends, but BMW's shaft drive made it trickier and more expensive to implement. But if the Führer wants a pair of rear shocks, you fit a pair of rear shocks.

pressed-steel frame design to the 750 twin class. The firm had struggled with the brazed steel-tube construction used previously, and some production inconsistencies led to cracking. These riveted pressed-steel frames were heavier and less elegant but easier to produce, with minimal skilled welding or brazing needed.

Further chassis developments arrived in 1935, with the first oil-damped telescopic front-fork suspension fitted to the R12 and R17 750s. The R12 tourer had a 20-bhp side-valve engine and the R17 a much sportier OHV design with 33 bhp. Both still used a rigid hardtail frame with no rear suspension at all. Indeed, it was another couple of years before BMW released a bike with full suspension; the plunger-type rear shock absorbers first seen on the 1937 race Kompressors made their way onto the road bikes for 1938—the 500cc R51, 600cc R61, and the two 750s, the R71 and R66.

A new 500 class engine also appeared: the 494cc twin-cam Boxer used in the 1936 R5. Each cylinder had its own camshaft, allowing shorter pushrods and higher efficiency. The R5 engine made around 24 bhp, a 6-bhp hike over the previous R57 500 sports Boxer, and had a foot-operated gearshift on the left-hand side for the first time. Added to a new steel-tube frame design and telescopic forks, it made for a surprisingly modern-looking design and indicated where BMW bike design was heading.

Finally, BMW's first 600cc machine made it into production just before World War II broke out and ended civilian motorcycle development. The R6 took the new frame design and foot gear change from the R5 and added it to a long-stroke side-valve 600cc engine, aimed at sidecar use. The R6 was soon replaced by the R61, which added a plunger rear suspension to the package.

Hitler Demands a Better Back End

According to BMW's own history, Adolf Hitler himself mocked the lack of rear suspension on the

1938–1940 R51

A landmark machine that took the excellent R5 engine and frame technology and added the new plunger rear suspension setup, the R51 was the first BMW road bike without a hardtail rear end. World War II prevented further development of this sporty little roadster, but it reappeared as the R51/2 in 1950 when BMW started postwar Boxer twin production.

The R5 engine found its natural home in the R51 chassis, complete with rear suspension. This cutaway picture shows the compact, clean design. Note the hairspring-type valve springs and timing chain. This drives both camshafts as well as the generator, all mounted above the crankshaft.

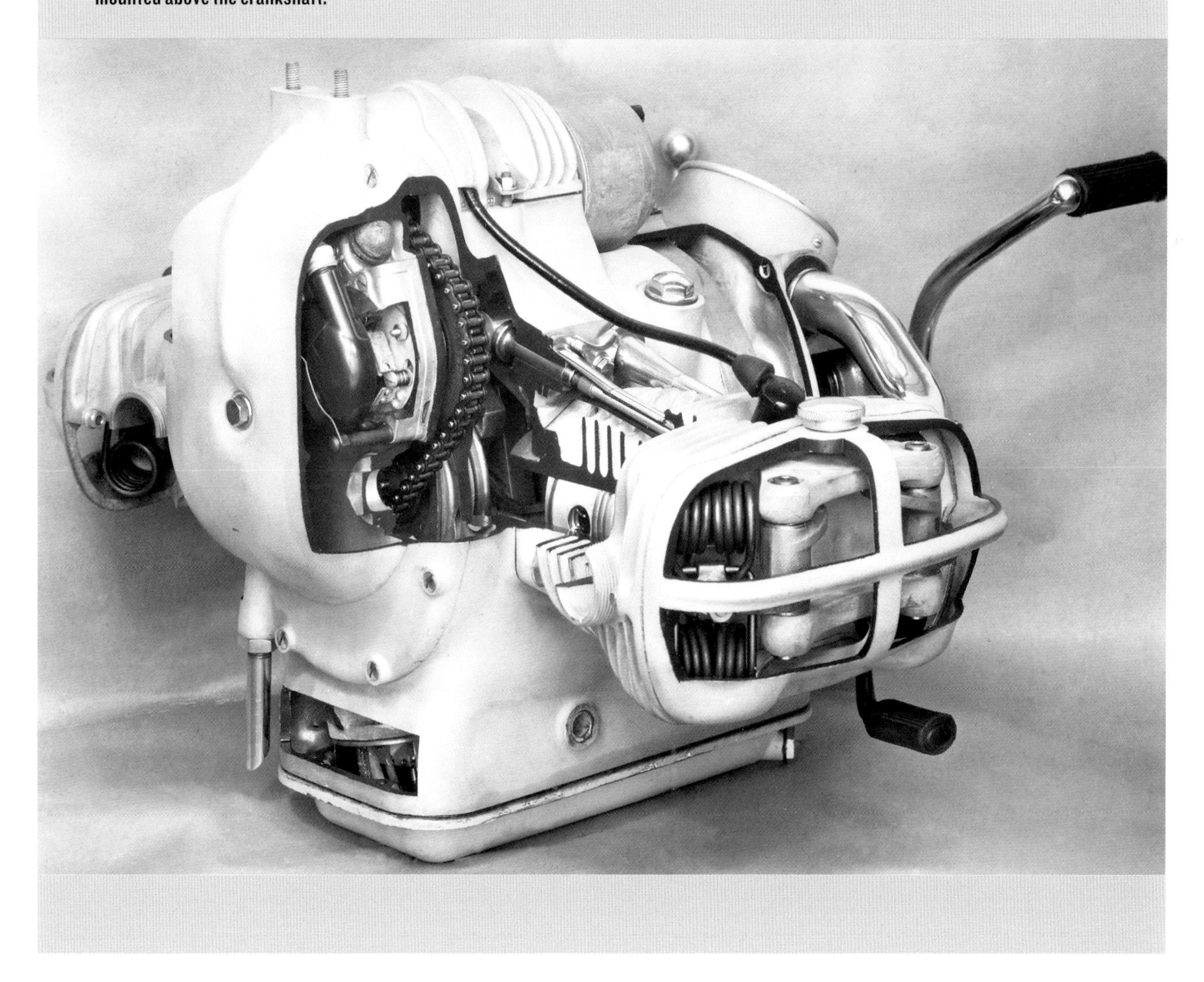

Shaft Final Drive

The rotary shaft is one of the most fundamental pieces of mechanical engineering ever invented. Modern motorbikes are packed with shafts, driving all sorts of parts inside the engine and within the chassis. But its use as a final drive is still fairly niche: the vast majority of bikes on the road today use a roller chain and sprockets to transfer power from the engine to the rear wheel. On the face of it, this seems ridiculous; running a long drive chain in an unsealed environment, close to the dirt, water, grit, and muck flung up from the road, wouldn't make it past the first design committee nowadays. Hundreds of small precision bearings, rollers, pins, and plates spinning at high speed, soaking up a grinding-paste mix of water and debris while transferring up to 200 bhp onto a skinny aluminum sprocket for thousands of miles? Madness, no matter how much lubricant the owner sprays on it when they remember.

ABOVE: On the early hardtail bikes, the shaft drive was a simple setup, with no expensive CV joints needed. The engine and wheel were both rigidly mounted, so there was no shaft movement to accommodate.

RIGHT: In 2020, BMW's M Endurance chain finally combined the best features of shaft and chain drive, with a diamond-like coating on a sealed O-ring chain. BMW claims that it needs no external lubrication yet eliminates wear, making it an essentially maintenance-free final drive.

Of course, it works. And a modern bike chain is an unsung wonder of moto-engineering: super-tough metals are able to cope with the conditions, and the invention of the sealed O-ring chain, where small rubber seals keep lube inside the rollers, was a game-changer in the 1980s. But back in 1923, none of this technology existed, and chains were weak, unreliable, and difficult to maintain, not to mention messy. Low-powered pioneer bikes had used belt drive, but that was no use for high-power outputs like the mighty 9 bhp of the BMW R32. And in the spirit of the high-quality, almost overengineered design ethos of the BMW machines, a well-designed shaft drive was the obvious answer. It fit perfectly with the transverse layout of the Boxer motor, allowing ideal cooling of both cylinders, with a simple transmission layout and a direct power transfer from the crankshaft to the rear wheel.

The R32 avoided some of the main engineering downsides of a shaft drive too: unsprung weight and the effect on suspension movement under power. On a modern bike, like the BMW R1250 GS, engineers design complex torque linkages to counteract the rise and fall of the rear suspension as the shaft transfers torque into and out of the rear hub. They also use lightweight aluminum alloys to reduce the mass of the shaft drive/swingarm

Paralever BMW R80 GS/R100 GS

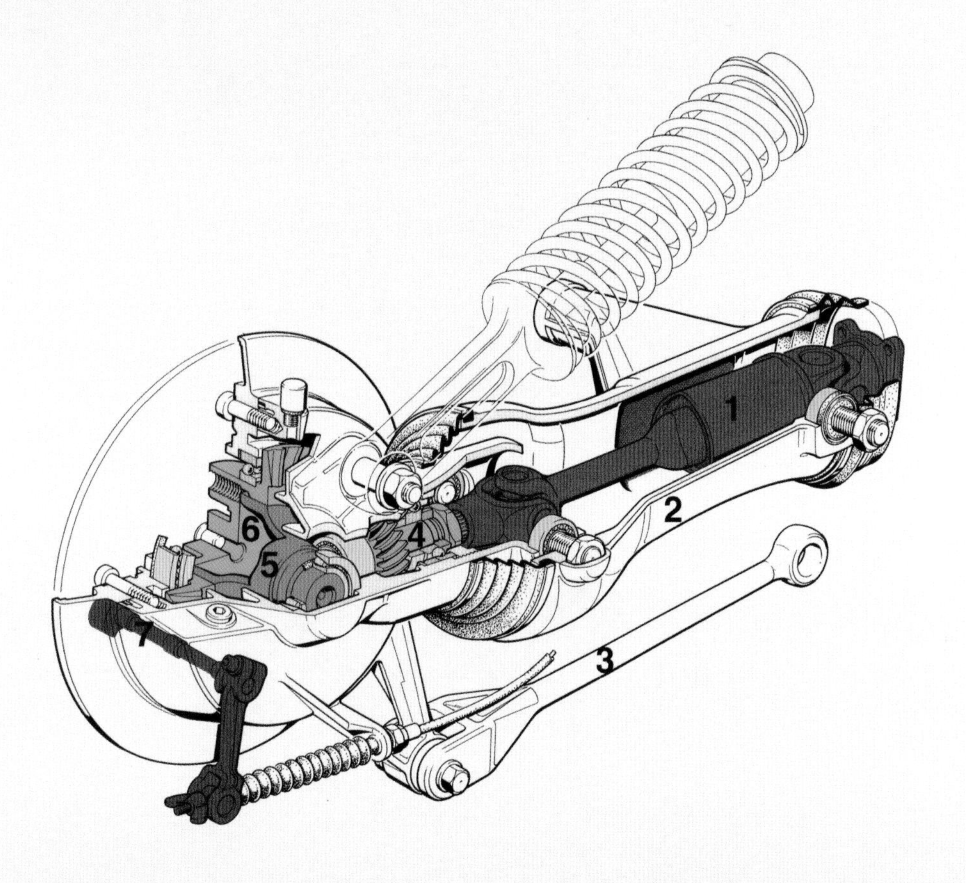

1 Doppelgelenkwelle mit Torsionsdämpfer
2 Doppelgelenkschwinge
3 Schubstange
4 Kegelrad
5 Aluminiumglocke
6 Tellerrad
7 Bremsschlüssel

BMW Motorrad GmbH + Co.
Kundendienst

Rear suspension means the shaft has to transmit power while the back wheel is moving up and down, so there are CV joints incorporated into the shaft design as well as linkages to counteract the effect of the shaft drive on the suspension. This diagram of the R100 GS Paralever shows shaft and CV joints in red. Note also the sliding spline joint at the wheel end to allow lateral movement.

components, cut the unsprung mass, and optimize suspension performance. The R32, of course, like all BMW bikes until the R51 of 1938, had no rear suspension at all. No rear suspension, no problems from a shaft drive interfering with rear suspension functions . . .

When rear suspension arrived, BMW stuck with the shaft drive and has been optimizing the technology ever since. Today's designs are extremely well developed, and careful computer-aided design has helped dial out much of the negative chassis effects of a shaft, especially on heavyweight machines.

BMW has also used chains on its bikes—the 1993 Funduro F650 was the first BMW without shaft final drive, and the firm has also relented on more lightweight bikes like the G310 range, where a heavy, expensive shaft wouldn't make sense.

Pragmatism also pipped tradition when the firm released its first superbike, the S1000 RR, in 2009. A proper racebike is impossible with a shaft drive: fast, simple final drive gearing changes needed at different circuits are totally impractical with a shaft setup, and the weight and torque effects are a massive handicap. BMW had a solid attempt with the R1200 S and the HP2 Sport variant, even running the one-make Boxer Cup race series. But against chain-driven competition, a shaft-drive superbike would have no chance on the track.

firm's bikes at the 1935 International Motor Show, a darkly comic moment from those terrible times. And in 1935 Germany, if Herr Hitler was mocking you, you did something about it—fast.

British bike companies like Norton were well ahead on this front, and the improved handling from having both front and rear suspension was undeniable. BMW engineer Rudolf Schleicher had been studying the technology and, together with another young designer, Alexander von Falkenhausen, came up with a "plunger"-type rear suspension setup that first appeared on the 1938 R51. The rear drive hub and wheel axle were suspended on sliding tubes, with springs inside to absorb bumps, giving a limited range of vertical movement for the back wheel. That movement meant the shaft final drive also needed a flexible joint, and while the early setups were fairly basic, it was another step forward in BMW chassis technology.

LEFT: Another legendary name in the pantheon of BMW engineers, Alexander von Falkenhausen helped Schleicher design the firm's first rear suspension setup as well as the high-performance R5. He also headed up the racing operation in the 1950s, overseeing BMW's massive success in the sidecar Grands Prix championship from 1954 to 1974.

OPPOSITE: Though he's shown here with an earlier R37, Rudolf Schleicher, together with Alexander von Falkenhausen, came up with the first plunger-style rear suspension design for BMW. The new setup, first used on the 1938 R51, was primitive compared with a proper swinging-arm design, but it was far better than nothing.

Single Life

At the lower end of the market, utility single-cylinder BMWs reappeared in 1931, four years after production of the first R39 single ended. The 198cc R2 qualified for a reduced tax rate, and riders didn't need a motorcycle license to ride one. Its 6-bhp output was well under that of the big twins, but the low price (995RM) meant it was a success, selling more than fifteen thousand between 1931 and 1937. Larger singles followed, the 398cc R4 (making 12 to 14 bhp) and the 305cc R3, with various levels of success. The R4 had enough power to compete with some 500cc machines, and it was popular among military, government, and police users as well as civilians. The R3, however, didn't have a big enough advantage over the R2 to make up for the tougher tax and license requirements, and it didn't sell well.

1932–1937 R4

Prewar, the peak of BMW's single-cylinder range was the R4: a 398cc OHV engine in a basic steel-tube chassis with the old-style leaf-spring front suspension, hardtail rear, hand gear change, and drum brakes at both ends. It had enough power—12 bhp—to compete with older twins and gained a reputation as a nearly indestructible workhorse. Together with the smaller 340cc R35, it was a lightweight (120kg), manageable solo machine for dispatch riders, reconnaissance scouts, and the like in the German Wehrmacht up to and through World War II.

World War II: Armageddon

By the summer of 1939, the writing was on the wall all across Europe. War was coming, and European nations were rearming in anticipation. The Nazi Führer, Adolf Hitler, had nothing less than world domination in mind and was building a war machine that would sweep across Europe in an unbelievably rapid lightning war, or Blitzkrieg. The Nazi war machine would chew up central and western Europe in 1939 and 1940, halting only at the English Channel and the North Atlantic coasts, before extending south to the Mediterranean and North Africa.

It then turned east in 1941 in an attempt to extend the German Reich to the Ural Mountains in the Asian part of the Soviet Union.

That gigantic Blitzkrieg war machine was made up of millions of men and millions of smaller machines. German industry was redirected into a total war economy unlike anything ever seen before. Firms like BMW halted all consumer production and were ordered to switch to armaments instead. Civilian motorcycle and car assembly was stopped completely in Munich and Eisenach, and the company's production facilities were turned over almost entirely to aircraft engine development and manufacture. That meant a massive ramping up of the firm's piston-engine powerplants, including the BMW 801 radial engine that powered the Focke-Wulf Fw190 fighter plane and the Junkers Ju88 transport, as well as a move into the new technology of jet engines. BMW took over the Bramo company, which had pioneered gas turbine development, and by the end of the war, it was building the BMW 003 axial jet engines in limited numbers for use on Heinkel and Arado warplanes.

Other German firms took over production of the cars, trucks, and other wheeled and tracked vehicles needed by the Wehrmacht, allowing BMW to concentrate on the airplane engine side. But BMW's motorcycles remained important to the war effort, and production of military two-wheelers continued in lighter solo form and heavyweight machines with a sidecar. The R12, with its virtually indestructible side-valve engine, was pressed into service as a sidecar outfit, while a range of smaller single-cylinder light and medium bikes transported dispatch riders, military police officers, and other itinerant state workers. Later on in the war, the more advanced R75 sidecar unit went into production, featuring a driven sidecar wheel for even better off-road performance.

BMW workers clear up bomb damage at the Munich aircraft engine factory following another Allied air raid.

Air power was vital to the German war effort, and BMW's 801 radial engine was one of the mainstays. BMW developed an analog computer to control the complex supercharger boost, ignition, and fueling settings, similar to how it would pioneer computer-controlled fuel injection in its bikes forty years later.

ABOVE: Germany was ahead of the game when it came to jet-powered planes: its Me 262 was the first operational jet fighter when it appeared in summer 1944. The 262 used BMW 003 jet engines in early prototypes, but development delays meant the alternative Junkers Jumo 004 engine was used in production. The BMW 003 eventually saw limited service in the Arado Ar 234—the first jet bomber—and the Heinkel He 162 emergency fighter. After the war ended, it was interest in BMW's jet-engine technology that attracted Allied forces to the Munich factory, and they confiscated everything they could find relating to the new tech.

The Dark Side of BMW's History

Like many large German corporations, BMW has a stain on its history from the Nazi period between the mid-1930s and the end of World War II in 1945. The total war economy ordered by Adolf Hitler, minister of armaments Albert Speer, and Erhard Milch, the state secretary at the Reich Ministry of Aviation, demanded unquestioning obedience, and nothing could get in the way of vital warplane engine production. And with a growing population in the prison camps built to house communists, homosexuals, trade unionists, and other "enemies" of the state, there was an easily tapped source of cheap labor. So the prison camps became *Arbeitslager*: labor camps.

As Germany invaded Czechoslovakia, Poland, the Low Countries, and France, the number of prisoners massively increased. By the summer of 1944, more than 7.5 million forced foreign workers were listed in German records, almost a fifth of the workforce. Prisoners of war, dissidents, and Jews—almost anyone from the occupied countries, in fact—were drafted to work as unpaid, forced/slave labor on construction projects, defense installations, and factories.

BMW wasn't the worst of the offenders from this time, but it did use slave labor in its factories. The Eisenach motorcycle production plant was around sixty miles from both the Buchenwald and

RIGHT: It's the most miserable episode in BMW's history: the use of forced labor, directed by the Nazi state, to produce aircraft engines. The firm has apologized repeatedly in recent years for these wartime activities, which saw civilian laborers from occupied countries, prisoners of war, and concentration camp inmates forced to work in its factories.

This BMW factory poster says "Carry your equipment into the shelter. It's too late when the bombs are falling! Important tools and equipment shouldn't fall victim to air raids, so let's get them to safety in time."

Dora-Mittelbau concentration camps, and Dachau (the first German concentration camp) was just ten miles from Lerchenauer Strasse, BMW's original base in Munich. Both Jewish slave labor and forced labor from other foreign workers were used to produce bikes and aircraft engines right up to the end of the war in May 1945.

Reconciliation and Contrition

BMW has since apologized for its role in the war crimes of Nazi Germany. Indeed, it was one of the first firms to address its history when it commissioned the excellent book *BMW, a German Story* in 1983. Written by Horst Mönnich, it laid bare the company's involvement with slave labor and the Holocaust, beginning a period of atonement and public debate over its past. Then, in 1999, the firm helped set up the Erinnerung, Verantwortung, Zukunft (Remembrance, Responsibility, Future) foundation to compensate victims of the forced labor program. Most recently in 2016, marking the one-hundredth anniversary of BMW engine production, the BMW Group issued a further apology, expressing "profound regret" for its activities during the Nazi period.

1935–1942 R12

The R12 first appeared in 1935, when civilian production was still important to BMW. Replacing the R11, it stuck with the heavier pressed-steel main frame and the side-valve 745cc engine design but added an extra gear to the transmission, now a four-speed. The R12 came in single and twin-carburetor versions, the twin-carb setup giving a few more brake horsepower but with more complexity and higher cost. There were 200mm drum brakes fore and aft and newfangled oil-damped telescopic forks up front, though the rear wheel was still rigidly mounted to the frame. The solo R12 weighed in at 185kg dry and could just about touch 70 mph with a fair wind.

Despite its basic rear end and modest power output, it was one of BMW's most successful models, and the firm built more than thirty-six thousand of them for civilian and military use, priced around 1,600RM each.

The side-valve R12 was a fairly sedate civilian machine before it was transformed into a sidecar-equipped military vehicle for the German Wehrmacht.

Sidecars

A motorcycle and sidecar combination is a fairly rare sight on the road today. But in the period between the wars, and from 1945 right up to the 1960s, they were common across Europe, providing families a practical way of transportation when a car was an impossibly expensive luxury. They ranged from simple home-built boxes on a rudimentary framework with a third wheel of some sort right up to huge coach-built behemoths with fully enclosed seating for two, plus space for luggage.

BMW incorporated sidecars into its motorcycle designs from the start, with options for lowered final drive gearing on models from the mid-1920s. Their slow-speed torquey power delivery and reliable, low-revving side-valve motors made long-stroke touring bikes ideal for use in combinations. The larger 750 Boxer engines from 1928 on were even more suitable for heavy-duty sidecar haulage. Later, in the late 1930s, the 745cc side-valve R12 was built in combination form for the army, and almost forty thousand were built between 1938 and 1941, equipping the German Afrika Korps during the North Africa campaign.

Perhaps the ultimate expression of the BMW sidecar first appeared during World War II, when the firm produced thousands of R75 sidecar units for the Wehrmacht between 1941 and 1944. Its 26-bhp, 745cc low-revving OHV engine was virtually indestructible, but it was the transmission that made the R75 such a workhorse. The final drive incorporated an extra powered shaft to drive the third sidecar wheel, and that, together with the super-low off-road gear ratios (and even reverse gears), kept the BMW outfit moving when lesser machines gave up. Clever design meant all three wheels were identical, so the spare carried on the sidecar could quickly swap in for any damaged rim.

For all their performance, the R75 and R12 sidecar combinations had some serious shortcomings, especially evident when the German army invaded the Soviet Union in June 1941 during Operation Barbarossa. The bikes themselves remained operational in terrible winter conditions but offered little in the way of protection from the elements (or enemy fire) for rider and passenger, especially compared with a half-track truck or armored vehicle. It was still better than walking, of course, and with a belt-fed machine gun mounted on the sidecar, a Wehrmacht BMW R75 was an extremely nasty thing to come across.

RIGHT: This R12 has been restored and is used by World War II reenactors. It's seen here at the War and Peace Revival show in Kent, UK, in 2010. ***John Goodman***

OPPOSITE: Another restored R12 is pictured in 2021 at the Military Odyssey show in Kent, UK. With a machine gun, seating for three, and plenty of storage space, it must have looked like luxury to a German infantryman on the Eastern Front in the winter of 1942–1943. ***John Goodman***

799 32
14 atü
935357

IIA-050

1941–1944 R75

In 1937, the German army's High Command wanted a more capable machine than the R12, and trials were held between BMW's new R75 prototype and the Zundapp KS750. Both were flat-twins with air-cooled engines, shaft drive, and a power takeoff for the third sidecar wheel. The trials specified a payload of 500kg—three beefy soldiers plus kit—and among other performance goals, the bike had to be able to travel at walking pace without overheating, for accompanying infantry on foot, and reach a 60-mph top speed.

BMW's flat-twin engine coped well with the slow-speed requirement, but the Zundapp won the trials in part thanks to cheaper, simpler construction. High Command actually asked BMW to start building the competitor machine instead of the R75, but the Munich firm was not at all keen. A compromise was reached by which Zundapp and BMW would attempt to standardize as many parts as possible between the two bikes, simplifying logistics and manufacturing.

The R75 had a new OHV engine design that featured several clever design tricks, like the large air filter above the fuel tank to mitigate dusty conditions in North Africa and the Russian steppe, a locking differential between the driven wheels, and reverse and low-ratio off-road gears. The brakes used 250mm drums all around, and the wheels were the same size as the German "jeep," the Kübelwagen. It would later become the basis for BMW's postwar production.

OPPOSITE: BMW engineer and R75 designer Alexander von Falkenhausen found himself researching his bike's performance in the most challenging environments—in the USSR on Germany's Eastern Front.

BELOW: Unlike the R12, the R75 was a completely new machine, with a unique transmission delivering power to the sidecar wheel too. I t was designed for hard-core military use, with heavy-duty components like the massive air filter inside the fuel tank area. The small helmet-shaped cover opens for easy filter maintenance. *Henry von Wartenberg*

ZH-27

4 Postwar and the 1950s

May 8, 1945, was Victory in Europe (VE) Day. Adolf Hitler had killed himself in his bunker on April 30, Soviet troops were fighting in the heart of Berlin, and Germany was in ruins. Hitler's successor, Großadmiral Karl Dönitz, authorized a total unconditional surrender to the invading Allied forces. The tragedy of World War II was over—in Europe, at least. Now, though, the work for peace had to begin. Most of Germany's industrial heartlands had been leveled, first by high-flying United States Army Air Force and Royal Air Force heavy bombers, then by Soviet assault guns. There was also widespread self-inflicted destruction caused by German SS troops, ordered by Hitler to impose a "scorched earth" policy. Factories, raw materials, food, and armaments were all to be destroyed rather than left for the forces invading from west and east.

Munich had been relatively secure for a while due to its location in the south of Germany, farthest from Allied air bases in England, and the BMW factories in the north of the city were still standing, though they'd taken their share of war damage. The BMW name was high on the agenda for the Allies, however, and not in a good way. The firm's main wartime activity—building warplane engines—put it in the spotlight for a number of reasons. First, the evidence of the use of slave labor in BMW factories and its part in the Holocaust against European Jews was becoming clear. Second, the firm's part in the development of high-performance piston engines had raised its profile as part of the Nazi war machine: plenty of Allied soldiers had heard the scream of a BMW 801 engine powering a Focke-Wulf Fw190 fighter-bomber as it attacked.

Finally, BMW's pioneering work on the development of jet engines marked it as a center of technical excellence. The results of this attention were twofold: the Allied military took all the secrets of BMW's engines, piston and jet, as well as what production facilities, machine tools, and assembly lines remained. There was also extensive looting, both from desperate civilians and released prisoners from the Dachau concentration camp. Then came an order from the Allied military government to level the main Munich factory buildings themselves.

The Eisenach factory, where motorcycle production had been moved after Munich concentrated on aircraft engines entirely, was first taken by the US Army in April 1945 before being handed to the Red

If you had the money (and the style), an R51/3 was ideal for sunny Sunday blasts around the foothills of the Alps south of Munich.

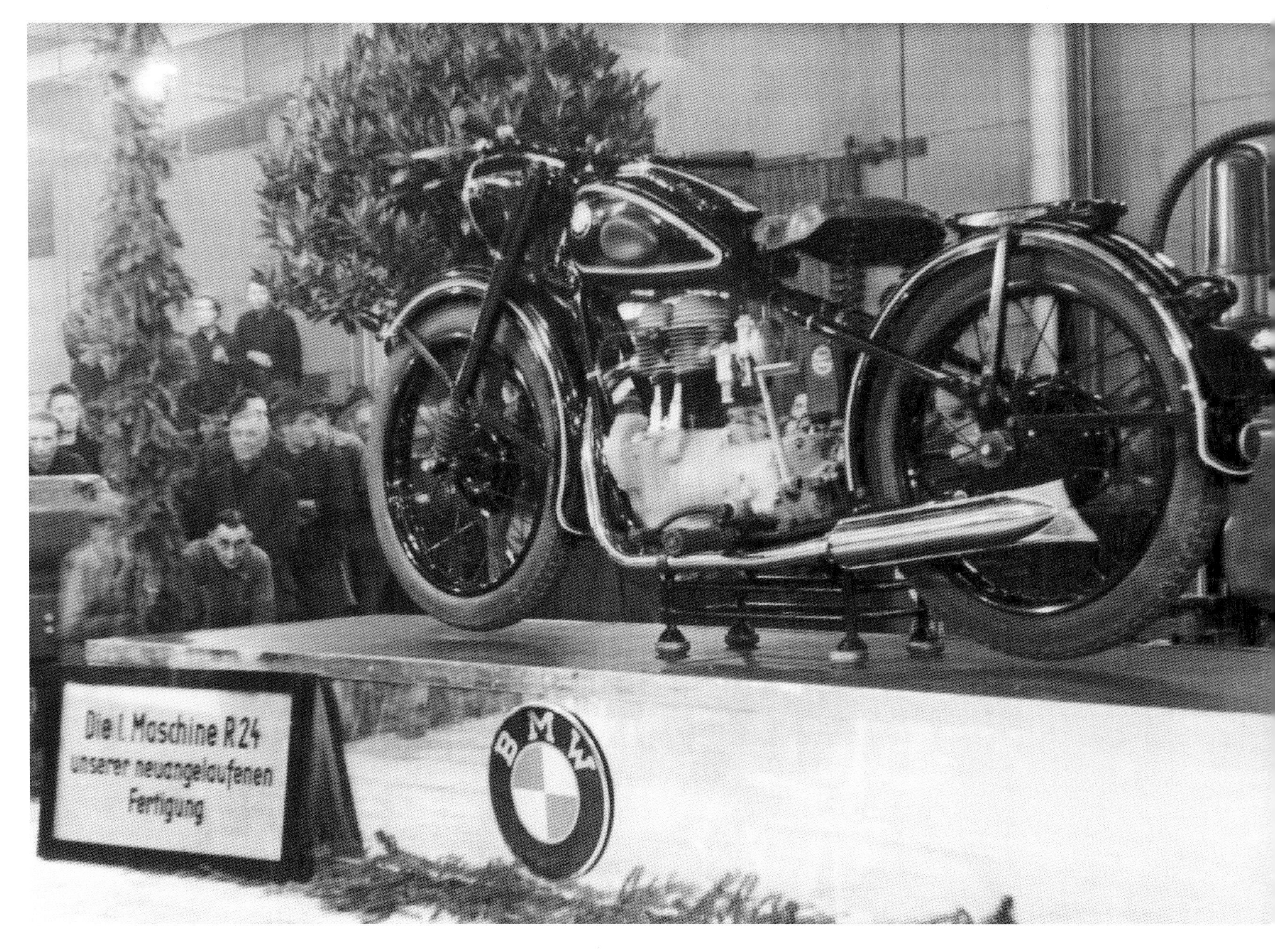

This R24 was the first new bike produced by BMW after the war. The sign reads "The first R24 machine from our newly started production."

Army a few months later in July. Later, in 1949, the Soviet Occupation Zone of Eastern Germany would be established as the German Democratic Republic, or DDR. Eisenach was just inside the border, and the BMW factory there was taken over by the state and rebranded as EMW—Eisenacher Motorenwerk.

Year Zero

By late 1945, you wouldn't have bet a lot on BMW surviving at all, much less returning to its former glory. The firm's factories in Berlin and Munich were wrecked, and the Eisenach plant, where it had moved all motorcycle production, now belonged to the Soviet Union. Germany was barely able to feed itself, much less build complex industrial products like aircraft engines. It was, in essence, a return to square one, a sort of "year zero" for the former industrial behemoth, and the initial years were incredibly tough for the ex-employees and managers in Munich. In the immediate term, a hand-to-mouth cottage industry of manufacturing sprang up, with small-scale production of essentials like cooking pots and pans. Stocks of aluminum left over from aircraft production were used to make basic household utensils, and the firm also designed and produced agricultural machinery as well as a rudimentary bicycle.

Back to Bikes

Car and aircraft production was still a long way off, and the obvious next step was the same as it was

1948–1950 R24

The first postwar BMW motorcycle was a reboot of the prewar R23. There was just one problem: all the plans had been lost. New blueprints had to be drawn up the hard way, by stripping and measuring an existing R23. BMW took the opportunity to upgrade the design. The gearbox gained an extra ratio and became a four-speed, though it kept the old-school manual gear lever on the right-hand side, as well as a left-side foot change. The 247cc engine had a new cylinder head that helped make 12-bhp peak power, 2 bhp more than the R23.

The chassis was mostly unchanged, with a rigid rear end, telescopic front forks, dual drum brakes, and interchangeable wheels. Perhaps surprisingly, it had lost a bit of weight and was 5kg lighter than before. Combined with the extra power, the R24 could now almost touch 60 mph.

The R24 was a big hit for BMW and played a large part in the company's resurgence from the ashes of war. More than nine thousand of the sturdy singles were sold in 1949 alone, at the princely sum of 1,750DM each.

ABOVE: Note the out-swinging kick start and battery box. Single carburetor and pushrod tubes are also on show.

LEFT: While big Boxers were out of the question, BMW was soon allowed to produce utility singles again after the war. This beautifully restored R24 shows how simple it was: sprung saddle, rigid rear end, basic short-travel front forks, and a shaft-drive single-cylinder motor. It was still a high-quality product, though, with deep chrome on the exhaust and coachlines and rubber pads on the tank.

1950–1954 R51 /2/3

As with the R24, BMW took a prewar design, the R51, and revamped it slightly for 1950s production as the /2 version. It used the same basic engine layout as before: 494cc capacity from a square 68x68mm bore and stroke with twin chain-driven camshafts. Indeed, the only obvious change was to the overhead valve covers, now a split design with a distinctive locating clip. Power output was 24 bhp (good for the time), and top speed was a heady 83 mph. The kick-starter shaft was, like all BMWs, mounted in line with the crank, so the lever swung out as you kicked, rather than back. That meant you had to get off the bike to start it, but luckily the Boxer motor was easy to start and idled well.

The chassis was very similar to the original R51, with the same steel-tube frame, plunger rear suspension, and telescopic forks. The wheels were a quick-release design, with 3.50 19 tires and 200mm drum brakes, though the braking performance was only just adequate according to magazine tests of the period. Equipment levels were high, with a locking tool box in the fuel tank and even an immobilizing lock: a simple steel locking bar that went through the wheel and frame, preventing casual theft.

BMW built five thousand R51/2 models before it was quickly updated to the R51/3 in 1951. The /3 version got an all-new gear-driven single-camshaft engine, though it kept the same 68mm bore and stroke and 494cc capacity. The valve covers went back to a one-piece design, and the alternator and Bosch ignition systems were relocated for a neater crankcase design. Fueling was by dual Bing carburetors, and the power output stayed around the same at 24 bhp.

The R51/3 chassis was largely unchanged from the /2, but later versions got uprated twin-leading-shoe drum brakes, then full-width drums. By 1954, the /3 had sold more than eighteen thousand units at 2,750DM—a massive boost to the firm's finances.

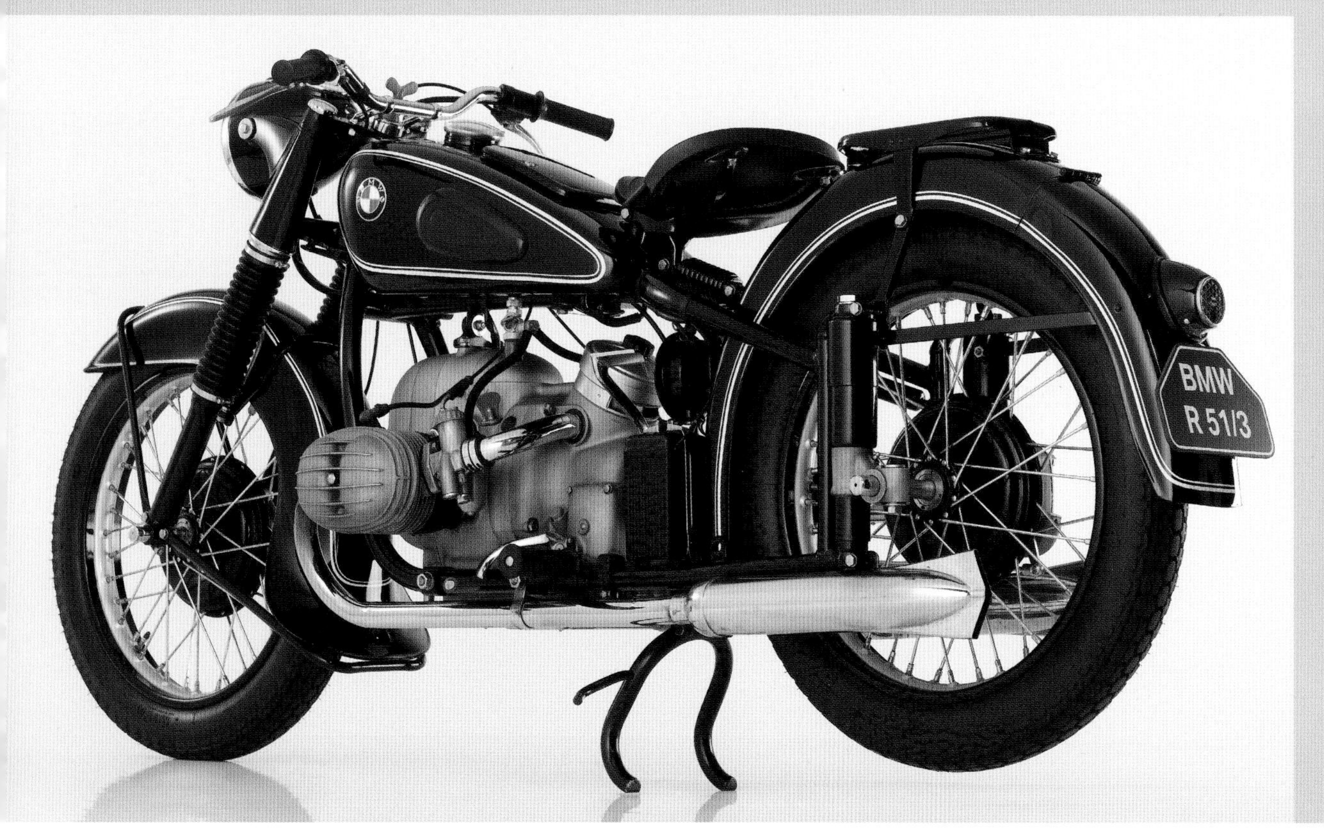

ABOVE: The R51/2 marked the return of BMW's Boxers to production in 1950. Paired with a high-quality sidecar like this Art Deco–styled Steib S350, it made a decent compromise between a basic motorcycle and hugely pricey automobile in economically tough times. *Henry von Wartenberg*

LEFT: This restored R51/3 shows off the final single-cam version of the R5-based engine, though the overall look is still a little dated. The plunger rear suspension and deep valanced mudguards both mark it out as a bike still based on prewar fundamentals.

The R67 was the second postwar BMW, a 600cc sidecar-lugger launched in 1951. It was updated to this /2 version in 1952, and the proper OHV engine, front and rear suspension, drum brakes on both ends, and the usual BMW quality made it a strong contender. *Henry von Wartenberg*

after World War I—motorbikes. This time, BMW had history and established designs, so getting back into the motorcycle game was arguably easier than starting out in 1923. The aim was simple: produce basic transport to help get a ruined nation moving again. Everything had to be approved by the occupying powers, however, and it was 1947 before American authorities allowed bike production to restart in Germany. The first postwar motorcycles would have just half the capacity of the R32 due to severe restrictions on the new machines. Engine size was limited to just 60cc at first, eventually rising to 250cc.

BMW engineers had toyed with the idea of a new 125cc two-stroke flat-twin, called the R10, and a prototype was produced. But when the higher 250cc capacity limit was announced, the prewar R23 247cc single was revived. This was no easy task—with motorcycle production lines moved to Eisenach and design blueprints either destroyed or seized by the Allies, the designers had to reverse-engineer existing machines to come up with new plans. The result was the 1948 R24, a 250 single with an updated engine design and four-speed gearbox bolted into a basic hardtail frame with telescopic forks.

The Boxer Returns

The capacity limits imposed by the Allied military government meant that BMW stuck to making its 250 singles for the rest of the 1940s. But by 1950, the rules were eased, and a 500cc Boxer twin was possible once more. The R51/2 was the result—a new workhorse machine that would sell by the thousands in /2 and /3 form between 1950 and 1954. BMW's first 600, the R67, arrived in 1951, featuring a large 594cc flat-twin long-stroke engine making 26 bhp. It was aimed at sidecar use and wasn't as popular as the 500 Boxers, though a /2 version of the R67 launched in 1952 and added better brakes and couple more bhp.

By the middle of the decade, BMW was well on its way back to health (it had also resumed car production in Munich), but there was a problem. The bikes being produced were still, largely, prewar designs, with a bit of tweaking here and there.

1950–1956 R25 /2 /3

By 1950, the R24's rigid rear end was looking rather old hat, and it was replaced on the R25 with a flashy new plunger suspension off the big Boxers. The engine was largely unchanged: the same inline-crank OHV single with 247cc capacity, a four-speed gearbox, and an updated shaft drive to deal with the plunger suspension. It was ready to use with a frame-mounted sidecar, and despite the high price (1,750DM), it sold like hot bratwurst. More than one hundred thousand R25s, /2 and /3 models, were sold between 1950 and 1956.

This 1955 R25/2 shows the "next-generation" single that replaced the R24. Plunger rear suspension, turn signals, and sprung dual seats mark this as a high-spec version, though the engine is basically the same 250 single as on the old R24. ***Henry von Wartenberg***

The large drum brakes and sleek chassis mark the 1952 R68 as a speed machine, but it still has the dated plunger rear suspension.

They were fine for the home market, but BMW's international competitors (mostly the British firms BSA, Norton, and Triumph) had more capable chassis designs, a wider range of machinery, and lower prices. If BMW was to compete, it had to overhaul its range of bikes, particularly in terms of frame and suspension technology.

BMW Chassis Tech Gets into Full Swing

For 1955, there was a big change in chassis tech across BMW's high-end range. The firm had made great advances in its engine tech, with the 1952 R68 making 35 bhp, enough to hit 100 mph. But it still used primitive running gear: plunger rear suspension and basic telescopic front forks. The handling had been overtaken by the power available, and something had to give.

The result was the so-called full-swinger BMW chassis design, first seen on the 1955 R50 and R69. Put simply, the new frames had a conventional swinging arm at the rear with twin shocks and an Earles-type front fork, with a leading link swingarm and twin shocks. This was a much stiffer front end than the spindly forks of before and was perfect for sidecar use, where the triangulated wheel mounting could better resist the cornering forces.

1955–1960 R50

The R50 appeared in 1950. It had a slightly more powerful 494cc Boxer twin motor thanks to a higher compression ratio and other mods compared with the R51/3. The big changes were on the chassis: a swingarm rear suspension setup, with hydraulically damped shocks and the shaft drive integrated into the right-hand side of the swingarm. The rear shocks used a similar mounting setup to the plunger units, so it could use the existing frame design with minimal modifications.

The R50 had a completely new front suspension setup too, with an Earles leading-link design replacing the old telescopic forks. It was the first of the "full-swing" BMW models and set the scene for the next fifteen years of BMW chassis design.

The R50 brought in all-new chassis technology, with Earles front forks and a sort of "halfway-house" rear swinging-arm suspension setup. Note the plunger-type top mounts for the rear shocks; this reduced the changes needed to the frame design, but it restricted wheel movement compared with properly engineered swingarm setups. The heavy construction of the Earles fork is also clear here—it adds a lot of mass and inertia to the steering.

1952–1954 R68

Launched with much fanfare in 1951 at the Frankfurt bike show, the R68 was a high-performance flagship sportster and claimed to be the first 100-mph production machine. Its 35-bhp, 594cc engine had a high compression ratio, big valves, and a race-spec camshaft. Aimed squarely at the high-power sportsbikes from the UK, the R68 could match them easily in terms of straight line speed, but its old-tech chassis, almost identical to the R67 sidecar lugger, let it down in the bends. Just 1,452 were sold, little surprise considering its hefty 3,950DM price tag.

RIGHT: BMW promoted the 35-bhp R68 as part of its new production methods. This 1952 ad showed off the new factory and the bike's claimed 160-kmh (99.4-mph) top speed.

FOLLOWING PAGES: There aren't many 1950s images as lively as this from the BMW archives. It shows Belgian journalist Marianne Weber thrashing an R68 on a road test for the French magazine *Motorcycle* in 1952. No helmet or any protective gear, but she managed to hit 99 mph on the road.

BMW's Rear Suspension Systems

A suspension setup for the rear wheel is ubiquitous in motorcycle design today and has been, pretty much, since World War II. The only bikes on the road without it are extreme hardtail custom machines, although some factory versions of these often have a "hidden" shock absorber for the hardtail look with a little more in terms of comfort and handling.

Pioneer motorcycle designs like the R32 did without rear suspension. The rear wheel mounted solidly in the frame, and only the pneumatic tire and sprung saddle isolated the rider from bumps in the road. The trailing rear wheel could deal with bumps at the modest speeds attained in these early days, but front suspension of some sort was essential. Hitting bumps at even low speed with a rigid front end will quickly lead to a loss of control and a crash, as well as being horrendously uncomfortable.

Racing, as always, drove improvements, as here rear suspension makes a big difference. Both tires need as much grip as possible to go fast, and there's absolutely zero grip when your solidly mounted rear wheel is kicked up in the air by bumps. A suspension system with shock-absorbing springs and a damping mechanism helps keep the tire in touch with the ground, increasing grip and boosting comfort and handling massively.

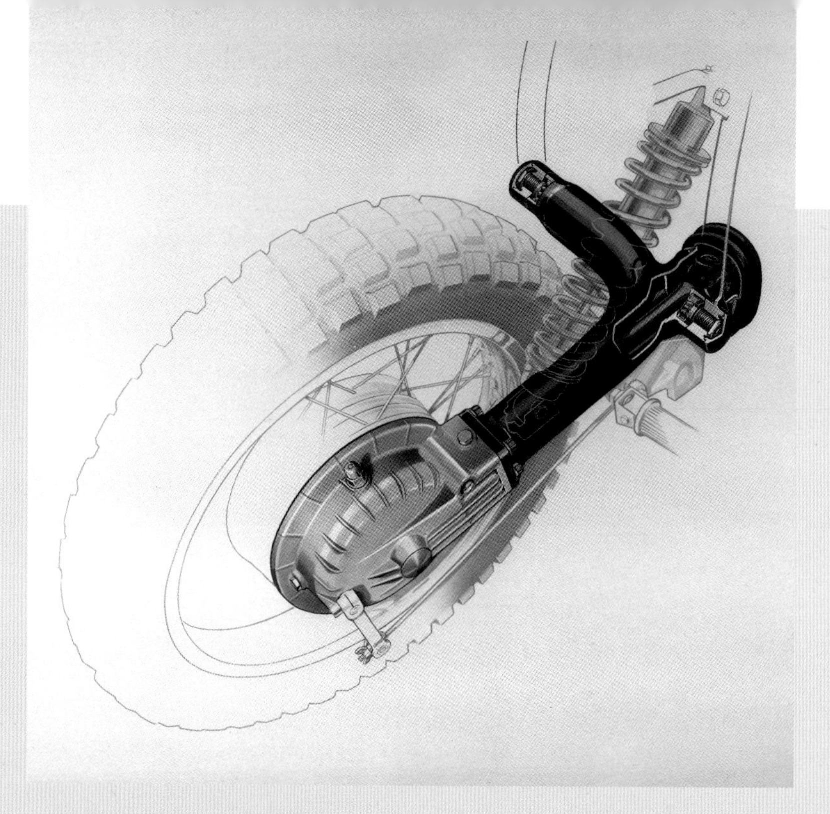

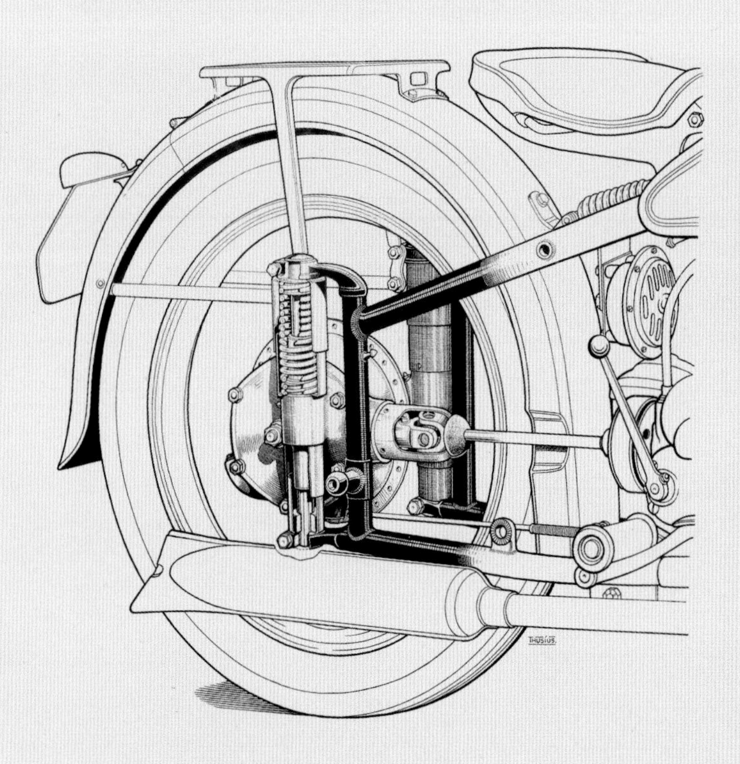

BMW was behind the curve in this area. By the mid-1930s, it was still using rigid rear ends, partly to reduce complexity on its shaft final drive. A solidly mounted engine and rear wheel only need a simple shaft to transfer power, while a moving, suspended rear wheel needs a universal joint arrangement so the shaft can follow the wheel up and down while the engine remains firmly in place.

The first BMW with rear suspension was the limited-production R51, which used a plunger-type setup. The rear wheel axle is mounted in a pair of sliding tubes, solidly mounted to the frame, with springs inside to absorb movement. When the wheel hits a bump, it moves up vertically within the plunger outer tubes, compressing the springs, then rebounds back down again, helping the tire follow the road surface more closely. Plunger suspension is far better than a rigid rear end, but it has its flaws. The amount of wheel movement is restricted, and the wheel can also twist in the frame since there's nothing to make the plungers on each side move by the same amount.

Swinging-arm rear suspension improves both of these problems: a rigid fork holds the rear wheel and rotates around its own axle mounted in the main frame. A pair of shock absorbers, with springs and damping mechanisms, controls the movement of the swinging arm and wheel while keeping the wheel perfectly in line as it moves up and down.

LEFT: This diagram of a 1938 R51 shows the plunger setup: the rear wheel axle is mounted on sprung housings, which can move up and down vertically inside a pair of cylinders. Wheel movement is limited, but it's better than nothing. A big downside is the lack of any lateral bracing, so the wheel can "twist" in the frame.

ABOVE: The R80 G/S was the first Monolever machine, and the single-sided arm was used across the range through the 1980s before being replaced by the Paralever toward the end of the decade.

The first BMWs with swingarm suspension were the exotic RS54 racebikes of the early 1950s. In 1955, the R50 saw a swingarm rear suspension system for the road. A steel-tube double-sided arm incorporated the shaft drive into the right-hand side, and a pair of hydraulic shocks absorbed the energy from bumps. Early designs were a compromise: BMW mounted the shocks in a similar way to plunger units, attached to the frame around the midpoint, which reduced the full travel potential. Later on, updated frame designs used a more conventional top mount for the shocks.

The shaft-drive twin-shock swingarm system worked well and looked after the back end of Boxers right up to the 1980s, when BMW's Monolever and Paralever systems appeared. With the Monolever, BMW worked out that with a massive stiff shaft housing on one side of its swingarm, it could dispense with the other side, fit just one shock, and save a load of weight. In the 1990s, a new Paralever back end gave much better handling thanks to a parallelogram linkage that reduced the impact of shaft drive forces on suspension movement.

TOP: Seen here on the R90 S is a conventional double-sided arm or fork that pivots at the bike end, with two shock absorbers on either side. This was the standard rear suspension setup on almost all road bikes in the 1960s and 1970s, but it wasn't ideal for BMW's shaft final drive. The back end would rise and fall under acceleration and deceleration, upsetting the handling.

ABOVE: A much more advanced setup, seen here on a 2004 R1200 GS, Paralever used a parallelogram linkage to reduce shaft interference with suspension movement. BMW has used the same principle on almost all of its shaft-drive bikes since 1987.

M-NR 5859

1955–1960 R69

Like the R50, the R69 was an update to the earlier 600 twin featuring the full-swing chassis: rear twin-shock swingarm suspension and front Earles forks. The long-stroke 594cc engine still made a healthy 35-bhp, but the R69 was more of a luxury tourer than a flat-out sports machine and made a great sidecar combination.

The R69 had largely the same "full-swing" chassis setup as the R50 but with the 35-bhp long-stroke 600cc engine. The fins on the OHV valve covers are an easy way to tell them apart: the R50 has five fins, the R69 two.

It was far from ideal on a solo machine. The Earles design is much heavier than normal telescopic forks, a downside both in terms of both overall and unsprung mass as well as steering inertia. An Earles front end made a bike heavier, with less precise front suspension movement, and required more effort to steer. In addition, the front-end lift under braking, together with the effects of the BMW shaft drive on the rear under acceleration, meant the chassis was never really settled during sporty riding on twisty roads. Nevertheless, BMW stuck with the full-swinger chassis design on the Boxer models right up to 1968.

OPPOSITE: The Isetta bubble car seems humorous now, but it was a deadly serious attempt by a troubled firm to survive.

The 1959 Crash

A decade on from the first postwar BMW bike going on sale, you might think things were going well. And up to the late 1950s they were, from a motorcycle point of view. Annual production at BMW had hit thirty thousand motorcycles by 1954. The next few years would see loads of new models selling for good prices, an all-new chassis design, and high-power Boxer engines that could match the best of the British bike industry.

But there was much more to BMW than just the cool bikes. Its car and aircraft engine divisions had been moving down their own pathways, but they were struggling. The aircraft engine division was sold off to the MAN company in 1955 as part of a complex deal to begin jet engine production for the German air force's Lockheed F104 Starfighters. And BMW's cars were faring poorly against competition from Mercedes in particular. The high-end cars were too expensive, and its low-end models, like the Isetta bubble car, made too little profit.

Then the motorcycle market imploded too. The German public was seduced by the growth of affordable small cars, which were much more useful than a 250cc BMW with a sidecar, especially in the winter. Annual sales of BMW bikes fell to 15,500 in 1956, and to a third of that in 1957.

Something had to give. BMW was losing money, and the sharks were circling. First American then British firms tried to buy the firm, before old rival Daimler-Benz proposed a merger, backed by BMW chairman Dr. Hans Feith at a company annual general meeting in December 1959. The planned merger failed, and BMW was saved by two half-brothers: Herbert and Harald Quandt. The Quandts were major financiers in Germany, and despite some qualms over their links to the Nazis (Harald's mother, Magda, had divorced his father and married Nazi propaganda minister Joseph Goebbels, committing suicide with him in Hitler's bunker in 1945), they expanded their shareholdings. The Quandts took charge, and BMW kept its independence.

Earles Forks

The first of a long line of "funny front ends" appeared on BMW bikes in 1955, a trend that has continued right up to today. The strange setup back then was the Earles fork, patented by British engineer Ernest Earles in 1953. It used a backward-sloping rigid fork attached to the steering head, with a small swingarm pivoting at the bottom end. A pair of shock absorbers mount between the forward end of the swingarm and the top of the rigid fork so the front wheel can move up and down over bumps, controlled by the swingarm and the shocks. As a result of the geometry of the Earles fork, the front of the bike would actually rise under braking, the opposite of what happens with "normal" forks.

The Earles setup was perfect for sidecar use: it was very stiff and stronger than telescopic forks, and the geometry could be adjusted easily by varying the swingarm pivot points. On a solo machine, though, it was too heavy and arguably held back BMW's chassis performance through the 1960s.

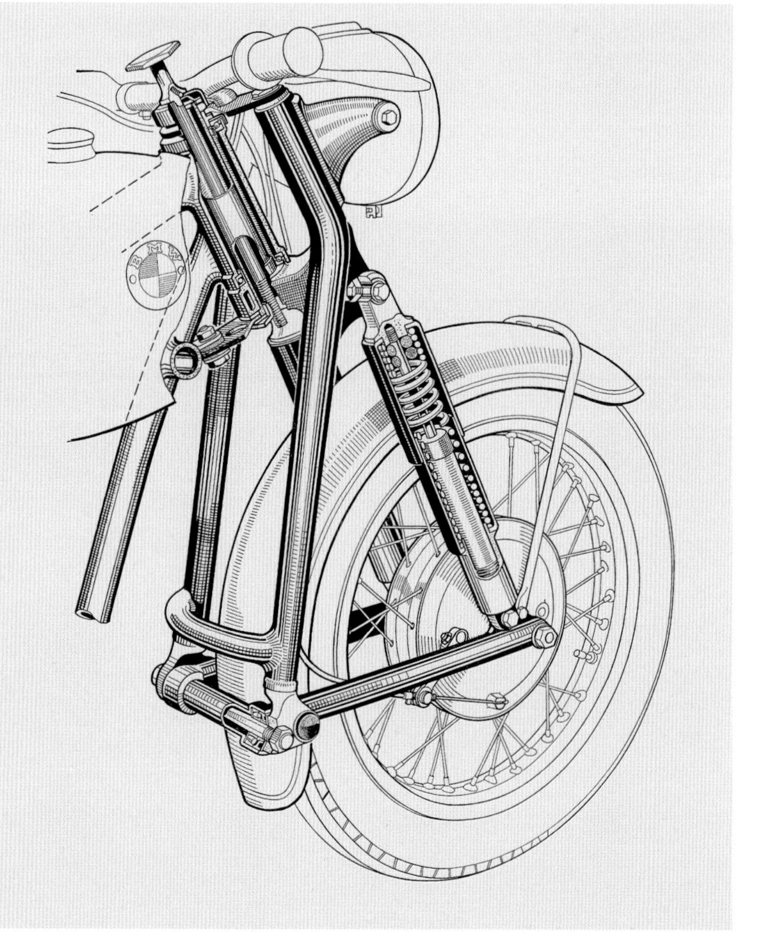

Seen here on an R50, the principles of the Earles fork are pretty obvious: a triangulated subframe houses a swinging arm with the wheel on the forward end, and twin shocks absorb bumps. The whole affair turns with the handlebars, so there's a huge amount of mass, which adds inertia to the steering. There's also too much unsprung weight for sporty handling. It's ideal for sidecar use, however.

1956–1960 R26

By 1956, the full-swing chassis had made it down to the 250 class, and the 247cc R26 gained rear swingarm suspension and an Earles fork. Having a chassis that was virtually the same as the R69 meant more weight and more expense, but it made for a sturdy small-bore sidecar unit. The motor now made 15 bhp, 2 bhp more than its R25/3 predecessor. Its top speed of nearly 80 mph made it one of the fastest 250s of the era. The R27 update in 1960 was to be the last single-cylinder BMW until the F650 in 1993.

RIGHT: BMW was justifiably proud of the big-bike chassis setup on its R26 single and advertised it as "built like a twin" in the United States.

BELOW: As with the plunger-suspension R25, BMW filtered the latest big-bike chassis tech down to its entry-level single in 1956. The R26 gained a full-swing chassis with Earles forks and semi-plunger rear swingarm suspension.

5

1960s and 1970s: Winning the Peace, Matching the Brits

The 1960s were a great time to be alive (or so my crazy old uncle always used to tell me). Rock and roll, miniskirts, giant flares, Woodstock, the Beatles, Hendrix, the Stones—incredible cultural moments and enormous global economic expansion. Postwar austerity was becoming a distant memory for many, and while there were plenty of downsides—the dark shadow of potential nuclear war, actual war in Vietnam, terrorism in the Middle East, bloody civil rights protests—it was, in general, a time of progress for humanity. For BMW in particular, the 1960s were tough. It had escaped financial oblivion at the start of the decade thanks to backing from the new major shareholders, the Quandt family, and investment was made in R&D and new products. These investments were mostly aimed at the car side of the business, however, where it seemed like the biggest profits would lie.

Motorcycles moved to the back burner somewhat, and looking back, a decision to halt two-wheeled development altogether would not have been surprising. The market was becoming increasingly difficult, with strong competition from the big British firms as well as the resurgent Italian bike industry. Norton, BSA, Triumph, and Moto Guzzi were building solid road bikes, while MV Agusta, Gilera, Ducati, and Laverda offered more exotic high-performance machines.

Meanwhile, Japan was stirring, its fledgling bike industry rising from the ashes of the war in the Pacific. Honda had been building small-capacity bikes since 1955, with some success, and Yamaha was actually now winning Grand Prix races with its two-stroke machines. Suzuki was competing at the front of TT races at the Isle of Man, while Kawasaki was selling a very decent 650 twin road bike, based on the British BSA A7, in 1965. Most other German bike manufacturers—DKW, Zundapp, and NSU—had fallen by the wayside, and with a promising range of cars now being developed in Munich, was there a future for motorcycles at BMW?

Luckily for all you BMW bike fans (and the author), motorcycle production continued. The

In the late 1960s, a high-end BMW like this R75/5 was a fabulous thing to ride.

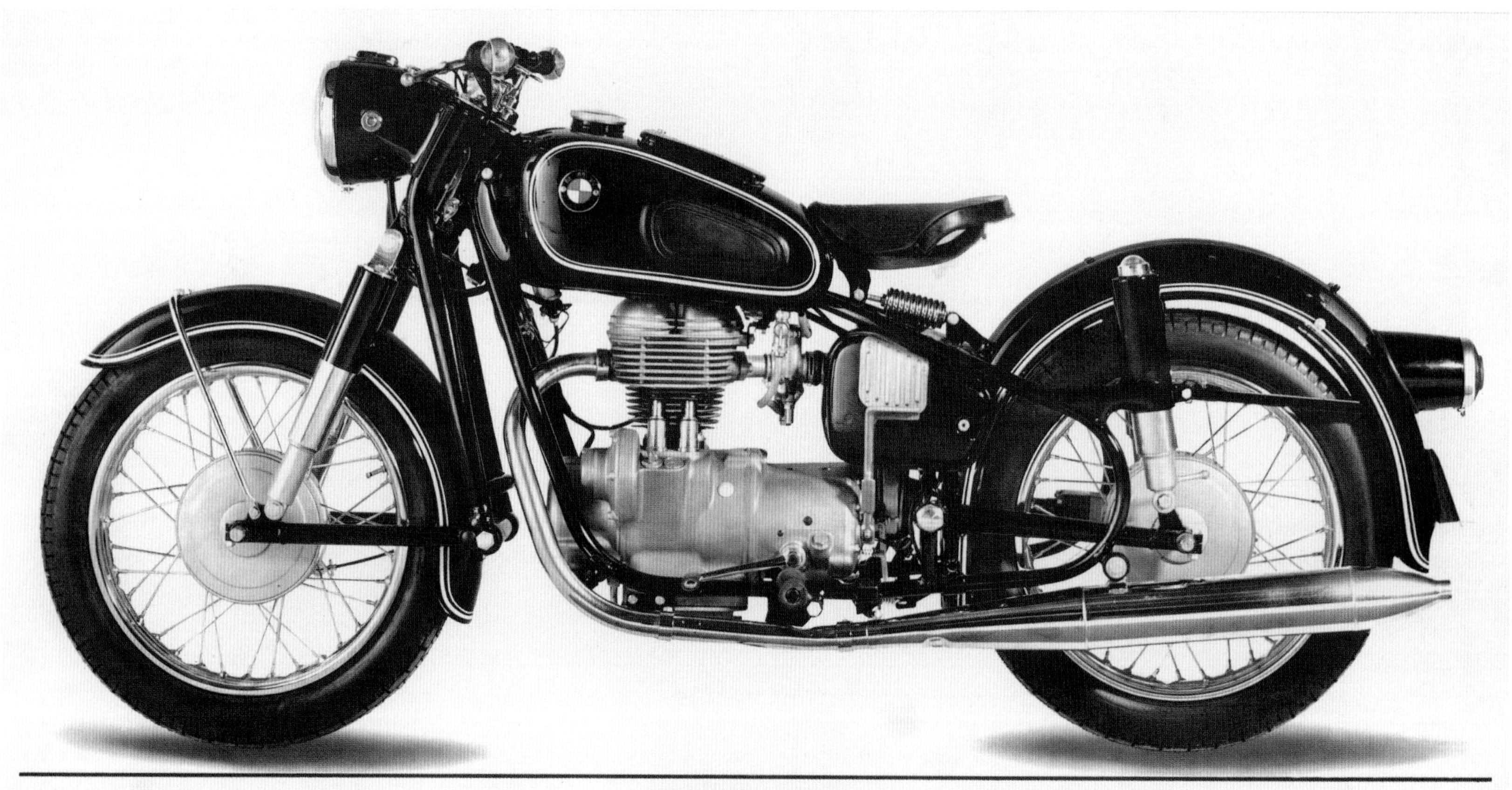

TOP: **The R27 was the last of the old-school BMW singles. After it was discontinued in 1966, the firm didn't make another single-cylinder bike until the F650 Funduro in 1993.**

ABOVE: **The old-style postwar Boxers were gradually being phased out in the early 1960s, but bikes like this R50 S were still an important part of BMW's range. Note the mid-mounted rear shocks, Earles fork, swing-out kick start, and sprung-style saddle.**

company stuck to the basics through most of the 1960s, with modest updates to the "full-swing" range of Earles-forked machines and tweaks to the last generation of the traditional Boxer engine design.

At the 1960 Frankfurt International Motorcycle Exhibition, BMW launched a new range that included the R50 S, R27, R50/2, and R60/2 models. These were all fairly minimal updates, but there was one other new bike there—the R69 S. It was a mix of old (such as the plunger-style rear shock mounting and 594cc engine) and new technology. The obvious change was to the seat, a new dual bench-style seat rather than the quaint sprung saddle designs. And for the first time, you could buy a BMW in a color that wasn't black. Careful readers might have noted that every bike pictured up until now has been black—because there was no other factory

1960–1969 R50/2 and R60/2

The last hurrah for the basic 494cc and 594cc Boxer engines that had kept BMW going for so long came in the form of the /2 versions of the R50 and R60 (though the R60/2 was mostly an internal company name). Both used the plunger-mount rear shocks and swingarm with the Earles front fork setup. The R50/2 engine now made 26 bhp, and the R60/2 produced 30 bhp. Both remained solid foundations for a sidecar combination, though sidecars became less common through the 1960s as cars continued to dominate.

Despite the old-school chassis and engine designs, both were the bedrock of BMW bike production in the 1960s, and the R50/2 sold more than nineteen thousand units, the most of any full-swing model.

BELOW: This 1962 R50/2 has a then-modern dual bench seat, which instantly makes it look much more modern than the old sprung-style saddles. A closer look reveals plunger-style shock mounts, Earles forks, and drum brakes.

RIGHT AND FOLLOWING PAGES: BMW has had extensive experience providing bikes for police and other emergency services over the past century. This police version of the R60/2 has the emergency lights and protective fairing, and it would also have carried panniers packed with radios and other kit. *Henry von Wartenberg*

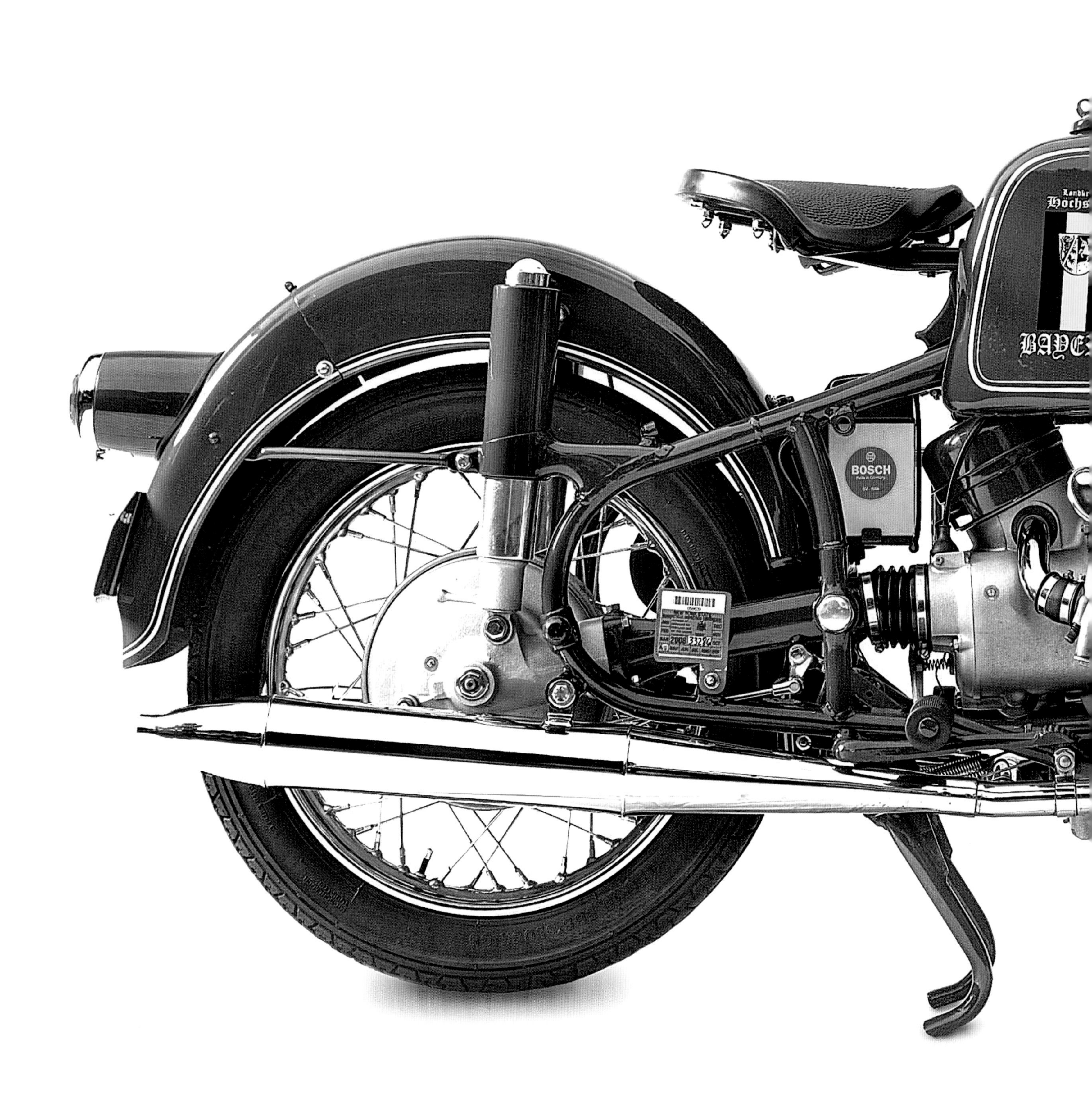
BOSCH

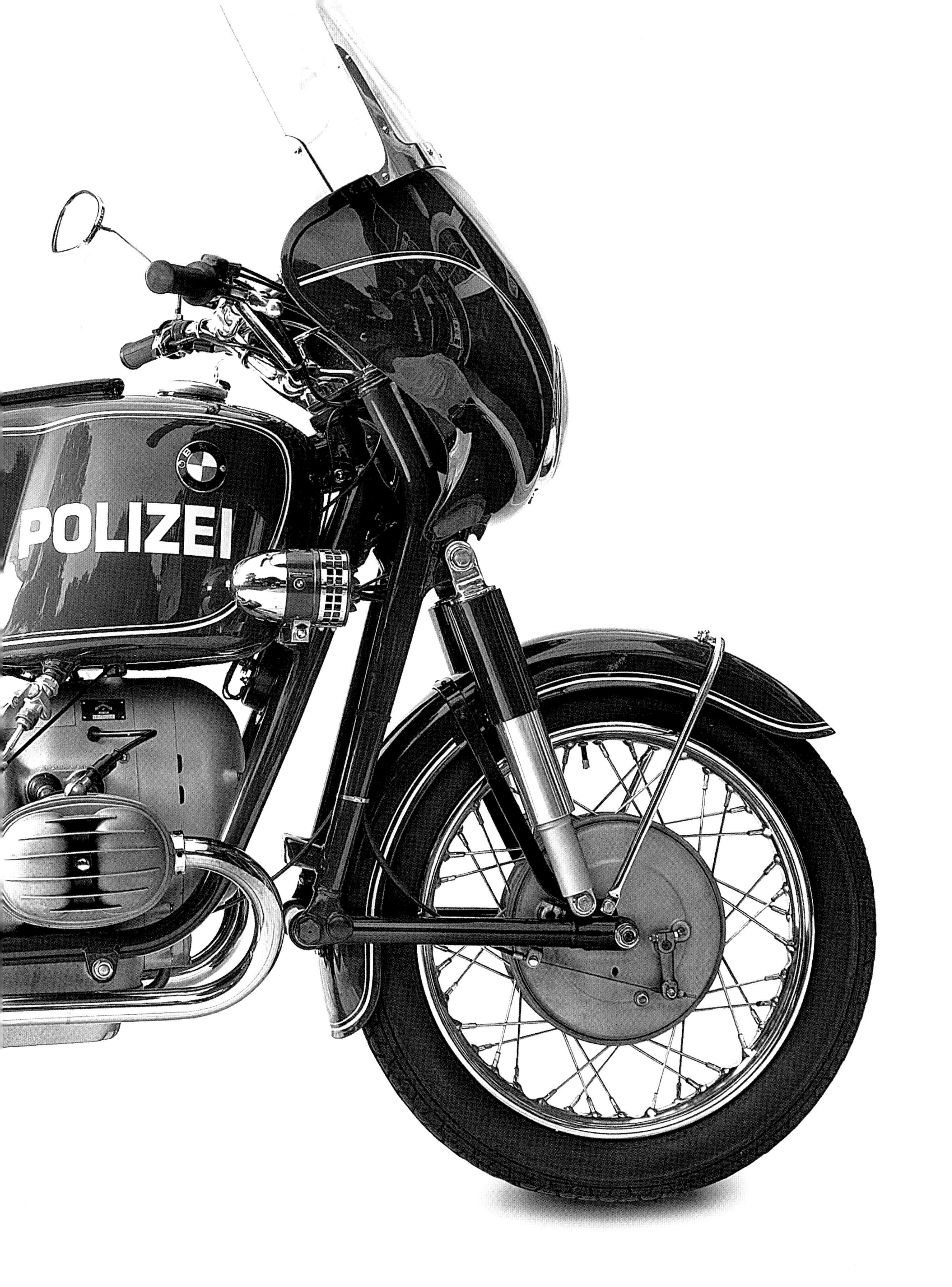
POLIZEI

The first bike built by BMW in Berlin was actually an R60/2 in 1967, made before full production of the /5 range began in 1969.

choice. Now, though, you could buy your R69 S in white from the dealer, and other colors soon followed: red, blue, green, and gray. Radical stuff.

What was really radical about the R69 S was its power and speed. A claimed 107 mph, from the highly tuned 42-bhp 594cc Boxer engine, was immense in 1960 and is getting toward modern middleweight performance.

US and Them

The last hurrah of the 1960s came from the United States, where customer demand drove the development of a couple of models specifically for that market. Ironically, perhaps, considering the modern tendency of the US market to favor long, stable, cruising machinery, American riders weren't happy with the sidecar-friendly Earles front end. So from 1967 to 1969, the R69 S, R50/2, and R60/2 were available in US versions, with a new telescopic front fork replacing the Earles swingarm up front. Luckily for the rest of the world, that wasn't the last time those forks would be used.

Into the 1970s: New Bikes Built in a New Factory

Today it's clear that by 1969, BMW needed to pull itself together a bit. Those pesky firms in Japan had moved on from making small-bore utility machinery and primitive two-strokes, and a veritable tsunami of incredible new bikes was on its way to Europe and the United States.

The first ripple from this massive wave had a Honda badge on it: the 1969 CB750. It brought almost unimaginable technology and performance as a 736cc inline-four-cylinder engine with a chain-driven overhead camshaft making nearly 70 bhp, twice the output of the R50 S. It came as standard with a front disc brake, 12-volt electrics, a push-button electric starter, five-speed gearbox—the list went on and on. It made everything else look a bit rubbish, as nothing from Britain, Italy, America, or Germany came close. And it would soon be followed by an even bigger shock in 1972—Kawasaki's 903cc Z1.

Suzuki and Yamaha soon joined in, and the big four Japanese companies put rocket boosters on motorcycle development throughout the 1970s. A slew of large-capacity, multicylinder two- and four-stroke (plus Wankel rotary) engine designs appeared, approaching 100-bhp peak outputs, matched later in the decade with chassis tech that could handle those levels of power. Dual disc brakes, monoshock rear suspension, advanced frame designs, aerodynamic bodywork, and more reliable electronics were all coming, fast. Unless BMW moved on from technology based on pre–World War II foundations, it was going to have a very tough time.

The Bavarian firm's answer came in 1969, with the launch of the "slash-five" range of bikes: the R50/5,

R50/5 and R60/5

The R50/5 and R60/5 were smaller 500cc and 600cc versions of the R75/5. They had basically the same running gear and engine bottom end but smaller bores and pistons. Power and torque were obviously down from the 750's 50 bhp, with 32 bhp from the 500 and 40 bhp from the 600. Weight was largely the same on the R60 as the R75, but the R50 weight was about four kilos lighter, mostly down to a missing electric starter on the 500. The smaller bikes both stuck with slide carburetors rather than the CV Bing units on the R75.

The R50/5 was rather underpowered for its weight and was dropped from the BMW range in 1973, when the /6 versions of the R60 and R75 arrived. Toward the end of the 1970s, the R60/7 appeared as the entry-level Boxer, with a single front brake disc, selling in small numbers before being replaced by the R65 in 1979.

The smaller versions of the /5 engine had the same massive bottom end, so the overall package was rather heavy for the power. On the upside, that also meant significant overengineering and longer life (in theory, at least).

BMW
BMW R60/5

A handsome beastie even now, the late-sixties **BMW Boxer /5** roadsters would fit right in outside the hippest cafés in London or Milan in 2023. Indeed, they're in strong demand from custom builders, for whom the overengineered motors, solid chassis, and classic styling are pure moto-catnip.

1960–1962 R50 S and 1960–1969 R69 S

Both S models used tuned versions of the base R50 and R69 engines but had bigger valves, increased carburetor bore sizes, and higher compression ratios (up to 9.2:1 on the R50 and 9.5:1 on the R69). The valve covers had just two ribs instead of the six ribs on the /2 machines, making identification easier. The extra performance caused some reliability issues on the R69 S; vibration at high rpm wrecked the crankshaft and destroyed some engines. An update rapidly appeared, featuring a damper unit on the front of the crankshaft to absorb the vibration and a new front crankcase cover to accommodate it.

A new hydraulic steering damper replaced the older friction design on both S bike chassis. It's important for modern riders to understand that back then, a steering damper was there to try to boost the stability of inadequate, weak frames and reduce wobbles and weaving at speed. That's quite different from today's bikes, where a steering damper is there to prevent tank-slappers when hard acceleration lifts the front wheel off the ground. No one was accidentally lifting the front wheel off the gas on a 35-bhp, 200kg R50 S, sadly.

Top speeds were considerable: 100 mph for the R50 S and 109 mph for the 42-bhp R69 S. The hefty weight and old-school full-swing chassis hampered handling, however, and these sporty S models were more grand-touring speed machines than super-sharp cornering weapons. The R50 S was sold for only a couple of years, and its total sale of 1,634 units makes it a fairly rare machine now.

ABOVE: The R69 S engine is probably the first motor we've come across in this book to offer anything like modern performance. Bigger valves and carbs plus high compression helped the 594cc unit produce a torquey 42 bhp.

OPPOSITE: A 200-kph speedometer was more a bragging rights device than a realistic indicator of top speed. The R69 S could get around toward 180 kph on there though. Note the large steering damper adjuster knob in the center of the steering stem.

VDO
km
20
40
60
80
100
120
140
180
200

BMW

S for sports—the R69 S was a real fire-breather for its day. Getting 42 bhp from the highly tuned 594cc Boxer meant it could top 105 mph easily. Note the "Denfeld" logo on the accessory dual bench seat.

R60/5, and R75/5. These were Berlin-built 500, 600, and 750cc machines, as the Bavarian factory in Munich refocused production entirely on cars (though bike R&D remained there). The Spandau plant in West Berlin became the heart of BMW motorcycle production, and remains so right up to today.

The new bikes all used a totally revised Boxer engine, with the same 70.6mm stroke and 498cc, 599cc, and 745cc capacities from 67mm, 73.5mm, and 82mm bores, respectively. An electric start came as stock on the 600 and 750 and was an option on the 500, requiring an all-new 12-volt electric system and powerful new alternator and battery to suit. This new engine architecture (retrospectively dubbed the "Airhead" by BMW fans) would be the basis for all BMW twins until the 1,085cc "Oilhead" four-valve Boxer appeared, twenty-four years later in 1993.

The all-new motor was bolted into a redesigned frame, with a steel-tube double cradle design, the telescopic forks first seen on the US models of 1968 and 1969, and a new mounting setup for the twin rear shocks and swingarm. The move from the plunger-style midmounts to conventional mounts at the top and bottom of the shock gave increased wheel travel and, together with the telescopic fork, much better handling.

Optimizing the Boxer

The /5 range held its own in the early part of the 1970s despite the onslaught of high-performance machines from Japan. BMW still offered something different, a more luxurious, high-quality "grand touring" philosophy compared with the performance focus of Japanese competition. How much of that was a realistic choice is moot, of course: it seems unlikely

Disc Brakes

BMW wasn't the first to fit disc brakes to a series-production road bike. That honor belongs to Honda and its 1969 CB750 four. But the Bavarian brand wasn't far behind: its R90 models came with 260mm front discs in 1973, a single rotor on the R90/6 and dual discs on the R90/S. BMW used a curious master cylinder arrangement; the handlebar lever actually operates a traditional Bowden brake cable, which is connected to a master cylinder under the fuel tank. At the time, BMW said this was to prevent vandalism and crash damage (though adding a cable to a hydraulic system seems like a very strange step now).

The front calipers were also unusual, with a swinging single-piston design from the ATE/Teves car components firm. Rather than having the caliper mounted on a sliding bracket like today's single-piston designs, the whole caliper rotated on a mounting pin when actuated, pulling the pad opposite the piston onto the disc's inner face.

The discs were drilled to reduce weight and improve cooling, and the overall system was far superior to the old-school drum brakes used before. BMW would stick to a drum for the rear, however. The lower performance was still sufficient for back wheel braking, and it was a simpler, cheaper installation than a rear disc.

The R90 S and R90/6 were the first BMW bikes with disc brakes, a dual disc on the S and single disc on the /6. Drilled rotors and single-piston swinging calipers were state of the art in the early 1970s.

If the R90 S hadn't been launched alongside it, the R90/6 would have been a very cool bike to ride in 1973. But with just one brake disc up front, no fairing, and "only" 60 bhp from the 898cc Boxer motor, it was very much the bridesmaid in the high-end BMW range.

that the German firm could have matched the likes of the Kawasaki H2 750 two-stroke, the Suzuki GS1000 four, or the mighty Honda CBX 1000 six with any air-cooled Boxer twin engine, the only engine tech open to it in the 1970s.

Nevertheless, BMW retained a loyal following and worked hard on the Boxer range through the decade. The "slash-six" range of roadsters debuted in 1973, and the R60/6 replaced the R50/5 as the entry-level machine. The R75/6 also appeared, sporting a smart new disc brake up front.

Bavaria had another trick up its sleeve for 1973—a pair of new 900cc Boxers: the R90/6 and the R90/S. A 90mm bore together with the 70.6mm stroke of the R75/5 gave 898cc capacity, making 60 bhp on the /6 and a massive 67 bhp on the /S. The Japanese superbikes were still pulling away (the four-cylinder 903cc Kawasaki Z1 was making 81 bhp), but BMW was still in touch, and the R90/S was a real connoisseurs' choice. It had almost as much straight line performance, but the comfort added by the small headlight fairing and luxurious seating made it a better long-distance choice. The big Boxer had the potential for mainstream solo superbike racing too, with impressive results in the United States; Reg Pridmore won BMW's first (and so far only) AMA Superbike title in 1976 on his heavily tuned machine.

The Beginning of the Modern BMW

The late 1970s saw a program of continuous updates to the Boxers, beginning with expansion to a full "liter-class" engine with the 980cc R100 series launched at the 1976 Cologne bike show. There were three R100 models: a base unfaired R100/7 with 60 bhp, an R100/S with 65 bhp (an update to the R90/S although with 2 bhp less),

The telescopic forks on this US model R69 S instantly transform the looks, replacing the massive, heavy Earles setup with a lighter, simpler front end. It still has the old plunger-mount rear shocks.

The Move to Spandau

BMW had owned a manufacturing facility in Berlin since 1939, when it acquired the factory as part of a deal with the Siemens & Halske/Bramo aircraft engine firm. At the end of World War II, the Berlin plant suffered the same fate as Munich; the Allies removed most of the machine plant, as well as blueprints, plans, and other material.

By the 1950s, the Spandau plant was up and running again as a component production site, making parts for both motorcycle and car production. This skilled workforce was thus in a good position to take on the full motorcycle production task in 1969, when the firm concentrated car production in Munich and moved bike production north to Berlin. R&D functions remained based in Bavaria.

The production move went extremely well, and by 1970, the factory was building more than one thousand bikes a month. Indeed, the Spandau plant built one hundred thousand bikes in just six years, with the landmark one hundred thousandth machine coming off the line in January 1975.

Today, more production facilities have been added in Thailand and Brazil, and the G310 single range is built by TVS in India, but Spandau in Berlin remains at the heart of BMW's motorcycle production.

TOP: This 1975 picture shows part of the original Spandau production facilities.

ABOVE RIGHT: Assembly line workers are regularly drafted in when landmark production figures are reached. This picture is from 1980, and Spandau actually passed its three millionth bike in 2019.

RIGHT: This computer-aided design image shows the 2001 update of the motorcycle production building.

1969–1973 R75/5, 1974–1976 R75/6, and 1977 R75/7

The flagship model in the new /5 range was the R75/5, a high-performance, high-tech machine with a 745cc version of the all-new engine. It produced 50 bhp, enough to hit 107 mph. Bore and stroke was 82x70.6mm, compression ratio was 9:1, and fueling was now by a pair of Bing CV constant velocity carburetors rather than the old-school slide carbs. CV carbs use intake pressure to automatically open the air slides to match gas flow into the engine, giving smoother acceleration and better economy at the expense of a little peak power production.

The 1969 Boxer design positioned the central camshaft below the crankshaft rather than above, driven by an automatically tensioned chain instead of a gear. The valve pushrod tubes were now hidden away underneath the cylinders, and there was a modern high-pressure pumped oil lubrication circuit. This came alongside a move from roller bearings to plain bearings on the one-piece crankshaft, using technology from the car division.

The gearbox stuck with four speeds (the later /6 update would provide a fifth), and an electric starter (also using car technology) came as standard for the first time.

The new bike had a fresh chassis layout based on a steel-tube double-cradle frame, telescopic forks, and twin-shock rear suspension, giving a curb weight around 210kg. The spoked wheels used a 19-inch front rim and 18-inch rear, with 3.25x19 and 4.00x18 Metzeler tires standard. The fuel tank held 24 liters (a smaller 18-liter tank was an option), and the electrics were now more reliable and powerful thanks to a 12-volt system and powerful 200-watt alternator.

The flagship R75/5 cost almost 5,000DM but sold more than thirty-eight thousand units from 1969 to 1973. The updated R75/6 in 1974 added a fifth gear and a front disc brake plus a host of other detail changes, while the final R75/7 variant only lasted a year before being replaced by the larger-capacity R80/7. The 750 class Boxer disappeared altogether.

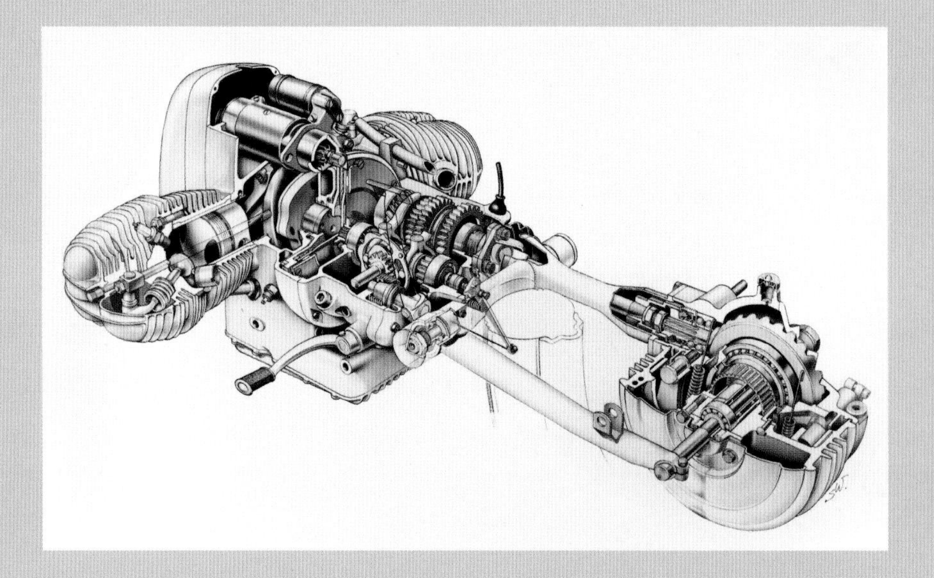

ABOVE: The /5 engine was very much a BMW Boxer—two valves per cylinder and air-cooled, with twin Bing carbs and a single cam under the crank. But the new design also featured a host of improvements and upgrades. The entire range used the same bottom end with a 70.6mm stroke, and different cylinder bores provided the various capacity options: 500, 600, and 750 initially. Electric start was available for the first time, and the Airhead engine would expand to a full 980cc and power a massive range of Boxers into the early 1980s and beyond.

LEFT: Finally, just in time for the 1970s, BMW had a new bike built in a new factory, with proper chassis tech and an all-new engine design. The R75/5 didn't have the snappiest name, but it led BMW's charge against the might of Japanese superbikes. The fundamental Airhead Boxer engine design used in the /5 machines would remain in production for nearly twenty-five years.

With 50 bhp and a sweet chassis design, the R75/5 was guaranteed to raise a grin on a twisty back road.

1973–1976 R90/6 and R90/S

Rumors of a 900cc Boxer had been swirling around the BMW cognoscenti for a while before the R90 machines finally arrived. At first glance, this was an unremarkable upgrade: the bottom end of the R75 engine seemed essentially unchanged, and a simple increase in cylinder bore size to 90mm took the capacity to 898cc. But BMW had gone further, making a series of strengthening mods to the crankcases, fitting a new crankshaft and alternator, and adding a five-speed gearbox. The top end saw bigger exhaust valves, new rocker bearings, and other detail mods to suit the more powerful motor.

There were two levels of tune: the base R90/6 made 60 bhp with a 9:1 compression ratio and 32mm Bing CV carbs, while the R90/S hit a mighty 67 bhp at 7,000 rpm, thanks to a higher 9.5:1 compression ratio and new Italian Dell'Orto 38mm slide carbs with accelerator pumps.

The chassis spec was broadly similar on the /6 and /S, with the same steel-tube cradle frame, shaft-drive swingarm, and Boge twin shocks and clocks. But there were some big obvious differences. The R90/S had a neat fork-mounted cockpit fairing that immediately marked it as something special. Designed by Hans Muth from BMW's car division (who would later design the legendary Suzuki GSX1100S Katana), it looked great and was also useful, boosting long-distance comfort by taking wind pressure off the rider at speed.

The /S also had a pair of front brake discs instead of the single disc on the /6. Those upgrades added a bit of weight: the R90/S weighed 215kg, 5kg more than the /6. But they made a massive difference to the style and performance of the /S, indicating something very new and very different from BMW.

All this fancy new kit came at a price, of course. The R90/S cost around 50 percent more than a Kawasaki Z1 of the same period, restricting it to wealthier superbike fans. BMW still sold more than seventeen thousand of them over three years.

OPPOSITE: This was BMW's big moment in the early 1970s. The flagship R90 S looked like a million bucks thanks to its Hans Muth design, small fork-mount fairing, and vibrant orange paintwork. Dual disc brakes and a tuned 67-bhp, 898cc version of the /5 engine made a potent riding package on road and track.

ABOVE: Rather commendably, BMW didn't shy away from showing women riders in its promotional imagery even in the early 1970s, when other firms were often a bit less enlightened.

RIGHT: A clock, battery gauge, 220-kph speedo and tacho, plus a slew of idiot lights made the R90 S cockpit more like a spaceship than a 1973 motorcycle. We still have the archaic steering damper knob though.

1976–1984 R100 RS

A full frame-mounted fairing, 70-bhp Boxer engine, and design by Hans Muth ticked all the boxes in 1976—there really was nothing like the R100 RS available anywhere else at the time. The Airhead Boxer engine got its ultimate bore expansion, up to 94mm, which gave a 980cc capacity with the same 70.6mm stroke inherited from the 1969 /5 range. A 9.5:1 compression ratio and 40mm Bing CV carburetors, together with bigger valves, rounded off the top-end upgrade, and there were other detail changes to the rest of the engine. On the chassis front, the steel-tube cradle frame had new, stiffer bracing, while the suspension and disc brakes were also tweaked from the R90/S setup. A 24-liter fuel tank and an extensive instrument panel made the RS a superb tourer. There was the option of wire-spoked or cast-aluminum wheels, and BMW's continuing upgrades to things like the halogen headlight made everyday life easier for the late-1970s biker. The price was still steep, around £3,000 (almost double the cost of a Kawasaki Z900), but even in the late 1980s, the R100 RS was a common sight on the road.

ABOVE: The R100 RS fairing was its trump card: a large frame-mounted unit that gave excellent wind and weather protection and also looked incredible for the time. It transformed the whole idea of what a high-performance touring motorcycle should be like.

BELOW: Building on the basis of the R90 S, the fully faired R100 RS was a huge leap forward in technology, design, and performance.

and the new flagship model, the R100 RS (RS in German stands for "Renn Sport" or Racing Sport). The most striking feature of the new RS was its frame-mounted full fairing—a first on a mainstream volume-production machine—together with a tuned 70-bhp version of the new 980cc engine.

Apart from the new liter-class bikes, 1977 saw mildly updated /7 versions of the smaller Boxers too: the R60/7 and R75/7. BMW also released the first in a long line of 800cc twins, the R80/7, which replaced the R75 from 1978 on.

The mighty R100 RS stole the headlines, of course, and it was a big hit for the firm for the rest of the decade, selling 33,648 units between 1976 and 1984. It completed the first steps taken by the R90/S and moved BMW bikes into a new realm of high-tech, luxury touring machinery, which BMW would make its own for the rest of the century. Looking at it more than forty years later, there's a clear lineage to BMW's current R1250 RS sport-touring Boxer via the 1980s K100/1100 RS, the 90s R1100 RS, the 2000s R1150 RS, and the R1200 RS of the 2010s.

ABOVE: 1976 saw another capacity boost to the R series with the 980cc R100 range. The R100/7 was the basic entry-level model, a naked roadster with a 60-bhp version of the new engine.

LEFT: The R90/S was replaced by the R100/S, which looked largely the same, lost a couple of bhp, and had to relinquish the limelight to the new R100 RS flagship model.

Aerodynamics

With its history in aircraft engineering, it's perhaps no surprise that BMW has a long association with aerodynamics and optimizing its bikes for speed. The firm set a number of speed records before World War II, assisted by streamlined fairings on its 750cc and 500cc supercharged machines. Later, with a growing car division that also required advanced aerodynamic design, BMW was able to justify extensive investment in wind-tunnel testing and the accompanying research program.

That led to the world's first production fork-mounted fairing on the legendary R90 S in 1973. Japan and Britain were lagging behind, with nothing in the way of faired road bikes, and BMW forged ahead, soon releasing the fully faired R100 RS in 1976. Wind and weather protection, together with improved high-speed stability and efficiency, became *de rigeur* on high-end machinery, and Japanese firms soon followed. By the 1990s, naked big-bore bikes were essentially a niche sector.

The next landmark was the radical K1 in 1991, which arguably took aerodynamics a bit too far for the market. Customers liked comfort and speed, but they also balked a little at the K1's extreme angular bodywork. Indeed, the aerodynamic design actually got in the way of the overall design: BMW fitted small storage spaces in the tail unit, so full-sized hard cases weren't an option. Matching soft bags were available but less than ideal in terms of security and weather protection compared with proper hard luggage, and the K1 suffered accordingly.

By the twenty-first century, aerodynamic design had softened a little, yet bikes like the K1200 S could look good and also have the wind-cheating abilities to easily cruise across Europe at three-figure speeds all day. The latest developments in motorcycle aerodynamics are pure performance. The 2021 M 1000 RR superbike comes complete with front-fairing-mounted carbon-fiber wings designed to produce downforce onto the front wheel at higher speeds, boosting stability and improving braking.

Small winglets on the front fairing of superbikes became fashionable after appearing in MotoGP in the early 2010s. BMW's first sportsbike winglets appeared on the 2021 M 1000 RR, and the firm claims that at 300 kph, the carbon-fiber aero devices generate 13.4kg of force on the front wheel and 2.9kg on the rear, for a total of 16.3kg downforce.

ABOVE: The fairing on the R100 RS was mostly for long-distance comfort and rider protection, but it also improved the aerodynamic performance of the bike at speed.

LEFT: This 1989 promotional photo shows off the land-speed-record streamliner bike, leading to the R90 S, R100 RS, K100 RS, and K1.

FOLLOWING PAGES: The author visited BMW's wind tunnel in 2014 as part of a presentation on riding kit. Besides testing cars and bikes, BMW also uses the facility to test helmets and jackets for things like noise and ventilation in a highly controlled environment.

GS

1978–1985 R45 and R65

They're a bit of a footnote in BMW history now, but at the end of the 1970s, the company put a fair bit of work into these small-bore Boxers. Unlike the rest of the range, the 473cc and 650cc engines didn't use the 70.6mm stroke bottom end pioneered in 1969. Rather, they had narrower, shorter-stroke (61.5mm) motors, with a different chassis setup better suited to the lower power outputs. The R45 was too slow, making just 35 bhp outside Germany. The home market had a low-power 27-bhp category, so the R45 made 8 bhp less there.

The R65 was a better proposition. It had a new Brembo front brake caliper rather than the old-tech ATE swinging design from the bigger bike. It made 45 bhp, on par with the likes of a contemporary Honda CX500, and weighed 205kg wet, which was lighter than the water-cooled CX500.

Neither were big sellers; most BMW customers wanted bigger bikes, and people buying smaller machines generally chose cheaper options from Japan. Nevertheless, they stayed in production into the mid-1980s.

The 45-bhp R65 offered much of the tech and design of the bigger BMW Boxer in a smaller, lighter package.

1000cc
K1
BMW

6

1980s: Trying to Keep Up with Japan

If, like me, you were born in the early 1970s, then the 1980s is a magical time in terms of motorcycle design. Even without the nostalgic rose-tinted glasses on, it was definitely an important period for bikes. Some of the most revolutionary designs first appeared in this period, bikes like Kawasaki's 1984 GPZ900R, which could top 150 mph on the road and filled the Isle of Man TT podium in its first year. Suzuki's GSX-R750 of 1985 brought the first true race-replica superbike to the road. In 1987, Honda launched its CBR600, which could win races as well as it could tour around Europe or serve as an unstoppable daily commuter. And with the 1987 FZR1000, Yamaha showed that a liter-class bike could still offer supersport handling prowess.

Japan was in its prime, with a stunning level of performance and technology and a punishing rate of development. Through the late 1980s and 1990s, the big four firms were generally unveiling drastic model upgrades every two years, with both engine and chassis performance improving (almost) every time.

In Germany, BMW was having to work hard to keep up. At the beginning of the 1980s, its engine range was restricted entirely to air-cooled Boxers, from 450cc to 1,000cc capacities. They all had two-valve heads with pushrod operation, big old Bing CV carburetors, five-speed gearboxes, and a basic, twin-shock rear suspension with shaft drive built into steel-tube swingarms. It was solid stuff and had a faithful following but was hardly cutting-edge.

Water-cooling, four-valve cylinder heads, aluminum frames, monoshock rear suspension, and other advances were all in the post from Japan, and again, BMW risked falling behind.

A New Engine for a New Decade

Big change was needed, and it came about in a very BMW fashion. A four-cylinder engine was required to make good power, but the idea of simply copying Japan with a standard transverse water-cooled engine, with the pistons and crankshaft arranged across the frame, was anathema to the Bavarian firm. A four-cylinder Boxer would maybe make sense, but Honda had already done this with its GL1000 Gold Wing in 1973.

Enter the K-series engine design: a fuel-injected, four-cylinder inline design but laid down with the cylinder head on the left-hand side and the crankshaft on the right. On the face of it, this design made little

The K1 design is futuristic even now—imagine how it looked back in the mid-1980s. The aggressive styling, wild aerodynamic bodywork, and brute-force engineering of the K-series engine felt like something from outer space.

TOP: The launch of the K100 range in the early 1980s promised a new high-tech dawn for BMW and an end of the old Boxer twins. But its customers had other ideas.

ABOVE: This 1982 picture shows the prototype K100 on the track. A massive preproduction silencer, flying brick engine layout, and single-sided rear swingarm are all on show, along with alarmingly minimal ground clearance.

sense and made for a wide, bulky bike, especially low down, restricting lean angles. But BMW had been making illogical motorbikes for decades, and just like Porsche with its rear-mounted, air-cooled sports car engines, it wasn't about to stop now.

The first K-series bikes were the three K100s launched in 1983 and 1984. The bikes used the same basic engine and chassis layout. The K100 was a standard naked machine without fairing, the RS soon followed with a slick half-fairing, and the RT completed the range in 1984 with a full-sized touring fairing and windshield. They instantly took BMW to another level of performance and technical complexity, but they also alienated some of the more "traditional" BMW fans, who valued the simplicity and ease of maintenance of the Boxers. The new K bikes were caught between a rock and a hard place—too "strange" for mainstream fans of Japanese machinery yet not "BMW" enough for the firm's hard-core fans. The performance was decent, sure, with 90 bhp from the K100 motor and a reasonable wet weight of around 240kg. But in the mid-1980s, the new BMW four-cylinder range was a bit of a curate's egg for the firm.

1983–1990 K100, 1983–1989 K100 RS, 1984–1989 K100 RT, and 1986–1991 K100 LT

In 1983, the first two K-series models launched: the naked K100 and the sport-touring K100 RS. Both shared the new 90-bhp, 987cc inline-four engine, five-speed gearbox, Monolever shaft drive, dual front disc brakes, and steel-tube bridge-type frame. The RS featured a sporty frame-mounted fairing with a large square headlamp. That fairing added 10kg to the wet weight; the RS was 249kg compared with the 239kg of the K100.

The following year saw another K100: the full touring RT, featuring larger full fairing and standard integrated hard panniers. The weight went up again, to 253kg, but the basic engine and chassis package remained the same. The RT was the last word in moto-luxury; it had ample accommodation for rider and passenger, an extensive dashboard, and useful touring add-ons like fairing storage pockets.

A couple of years later came the 1987 K100 LT luxury tourer. It was based heavily on the RT but added a standard-fit top-box and passenger backrest and even more touring equipment, including a Clarion radio/cassette deck and stereo sound system. Weight was up again to 283kg dry, while the power stayed at 90 bhp.

LEFT: The engine, transmission, and frame design are laid bare in this advertising spread, which focused on the ease of access to the new engine layout.

BELOW: BMW expanded the touring abilities of the K range to new heights with the 1984 K100 RT.

BOTTOM: The LT version of the K100 was aimed at mega-tourers like Honda's Gold Wing, with luxury accommodation for rider and passenger, plenty of luggage space, and a slew of optional accessories.

The K-Series Powertrain

There aren't many unique production motorcycle engine layouts—most have been used by several manufacturers—but the BMW K-series motor really is in a class of its own. BMW first built a prototype four-cylinder shaft-driven design using a small car engine (the "Suitcase" Douvrin engine from French carmaker PSA), in a laid-down setup, which it then developed into the 1983 K100.

The all-aluminum engine was arranged with the pistons moving across the frame, driving a crankshaft in front of the rider's right boot, with the cylinder head in front of the rider's left footrest. The intake manifold and Bosch Jetronic fuel-injection system were mounted on top, an airbox was under the fuel tank, and the exhaust pipes were routed vertically down before curving back toward the rear of the bike. A car-type dry clutch mated the engine to the gearbox, and power was then delivered in a straight line to the shaft

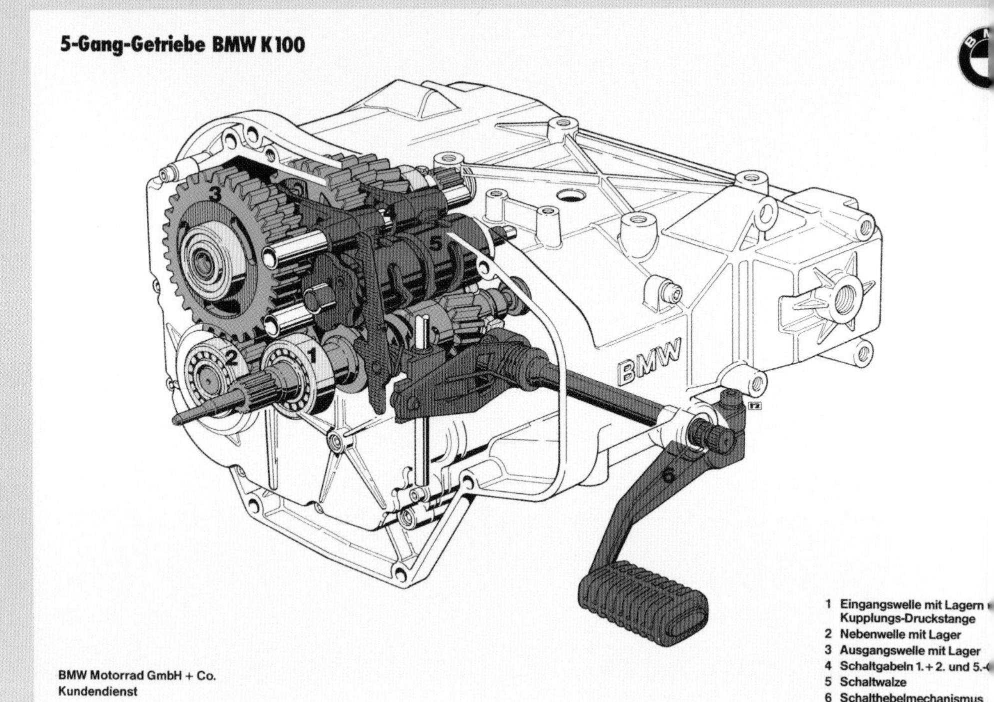

ABOVE: Separate five-speed gearbox and shaft final drive

RIGHT: The sixteen-valve version of the K100 engine, first used on the K1, stripped down to show off the high-performance internal parts. Note the narrow cylinder bore required by the laid-down engine design.

final drive, incorporated into a single-sided swingarm with direct-operating monoshock (the Monolever from the R80 G/S).

This fundamental design had a number of downsides, mostly due to packaging. Make the engine too "wide" from the top of the cam cover to the bottom of the crankcases, and you limit cornering potential. Make it too "long" from the front face of the engine back to the gearbox, and you end up with a very long wheelbase, which reduces agility. The original K100 had a long-stroke design partly to guard against this, using a narrow 67mm bore and long 70mm stroke—the complete opposite of what the performance-oriented Japanese engineers were doing. Bigger bores and shorter strokes are better for high-rpm operation since average piston speeds are lower (the piston has less distance to travel for a given rpm) and there is more space in the combustion chamber for big valves.

The original K100 unit made 90 bhp and had a 987cc capacity, two valves per cylinder, and a fairly high compression ratio of 10.2:1. The two overhead camshafts were driven by a single chain, and the engine used water cooling. BMW used then-novel electronic fuel injection on the K100, a version of the BMW Jetronic analog system used on its cars. Later, a three-cylinder K75 750 appeared, and the engine then gained a sixteen-valve head in 1988 before growing in capacity to 1,092cc in the 1991 K1100 LT, then up to 1,171cc for the K1200 RS in 1996.

One advantage of the K engine should be a low center of gravity—though the routing of the exhaust and the need for reasonable cornering ground clearance meant the motor actually needed to be mounted fairly high up in the chassis. Plus, a low center of gravity makes for a less agile bike in terms of handling. Despite the fundamental handicaps of the layout, BMW went all in on the K bikes for much of the 1980s. It then brought back the old-tech air-cooled Boxers as a stopgap while the next generation of oil-cooled four-valve Boxer twins was developed.

Most of the K-series machines worked well, especially the touring models. By the late 1990s, though, the sight of the whale-like K1200 RS valiantly claiming to be a superbike despite its massive weight was a little bit sad. The last hurrah for the laid-down K series was the luxo-touring K1200 LT, which was launched in 1999 and stuck around until 2010. By the mid-2000s, BMW had seen sense and launched a new K1200 model with a "normal" across-the-frame inline-four engine. It is tempting to wonder how the 1980s would have gone for BMW if it had resisted the urge to be different for the sake of being different and built a more traditional four-cylinder engine design from the beginning.

ABOVE: The wide, low K-series engine layout is obvious in this hard-cornering pic from the K100 launch in 1983.

LEFT: The original eight-valve K100 engine from 1983

ABOVE: **This is how the K series should have been from the start (arguably). The 2004 K1200 S used a "proper" across-the-frame inline-four-cylinder engine layout.**

LEFT: **The initial 1983 launch of the K series saw two new models: the K100 and K100 RS. Both used the same basic engine and chassis, in naked roadster form on the standard K100 and with a sporty-touring fairing on the RS.**

The Beginning of an Era: The First BMW GS

BELOW: This 1979 picture shows the first prototype of what would become BMW's most successful model range, the R80 G/S. The high-level exhaust, long-travel suspension, and dirt tires were all similar to what BMW racers and engineers were using on their own home-brewed dirt racers.

OPPOSITE: BMW's legendary Paris–Dakar racer Gaston Rahier is shown here on a bespoke racing R80 G/S prepared by the HPN (Halbfeld Pepperl Neher) chassis tuning firm.

The K series was soaking up a lot of energy for BMW in terms of R&D budgets and marketing efforts. Ironically, the seeds of its biggest two-wheeled success story so far were actually being sown elsewhere. BMW had a strong history of off-road race performance in its early years, but that had fallen away in the 1960s, and the firm didn't really have a competitive off-road machine. In the early 1970s, one of its engineers, Rüdiger Gutsche, had competed in off-road races like the ISDT (International Six Day Trial) with his own modified R75/5. Gutsche had realized that while the Boxer was much heavier than "proper" off-road bikes, the fact was that you didn't need a really lightweight machine for maybe 95 percent of off-road riding. Chassis mods for more ground clearance, a high-level exhaust, and a change to a 21-inch front wheel and dirt tires could go a long way, and Gutsche's bored-out 898cc R75 was far more capable than anyone expected.

In 1980 then, BMW decided to take a chance on a niche within a niche. The Paris–Dakar Rally had been launched a couple of years earlier as a hard-core off-road race for cars and bikes through North Africa. BMW had been part of the competition from the early days, when pioneering privateers like Jean-Claude Morellet competed in the 1979 Rally, and it would go on to launch full factory efforts and win the overall race in 1981, 1983, 1984, and 1985 with French rider Hubert Auriol and Belgian Gaston Rahier.

BMW
BMW
R 80 G/S
BING

1980 R80 G/S

It might have been dubbed Gelände/Strasse (off-road/road), but the secret of the R80 G/S's success was the same as it has been on every other BMW GS machine since. It simply was a very good road bike, despite its dirt-friendly design. Sure, you could ride one across the Sahara Desert with a bit of prep and a big dose of Bavarian bravado. But even if you never tackled anything scarier than a gravel car park, the G/S was a delight.

The new 247 E engine design was based on the existing R80/7 motor. Upgrades included new Nikasil-coated cylinder bore technology, contactless electronic ignition, Bing CV carburetors, and a lighter flywheel. On the chassis front, the frame came from the middleweight R65 and was fitted with the new Monolever rear suspension setup. This used a single-sided swingarm with the shaft drive and sole shock absorber on the right-hand side, with the rear wire-spoked wheel secured by three bolts. BMW claimed the single-sided arm was lighter and just as stiff as a double-sided design, and there was the added bonus of a quick-release wheel for easy tire/wheel maintenance.

The engine performance is sedate from today's point of view: 50 bhp from an 800cc bike is fairly modest. The weight would actually pass muster for a modern rider. A 186kg wet mass is not unreasonable and is actually 43kg lighter than the current (admittedly twice as powerful and packed with technology) F850 GS.

The R80 G/S was replaced by the completely overhauled R80 GS in 1987, which swapped the Monolever back end for BMW's new Paralever rear suspension system. The Paralever used a parallelogram-type linkage to isolate suspension movement from the shaft drive forces, canceling out some of the handling downsides compared with chain drive.

OPPOSITE: The first outing for a single-sided rear swingarm on a BMW. The extra strength of the shaft drive housing let engineers drop the other side, saving weight and making rear wheel changes far easier.

BELOW: The production R80 G/S launched in 1980 was much more polished than the modified dirt racers used by BMW enthusiasts off-road. But it still looks pretty hard-core to modern eyes.

Twenty years after the /5 Airhead Boxer engine launched, it was still powering new BMWs like this 1989 R100 RS. Suspension and brake upgrades from the K series plus an R80 G/S Monolever back end sharpened the chassis a treat, but the 980cc engine now made just 60 bhp.

Together with the experimental off-road racebikes built by enthusiastic racer-engineers, there was suddenly a basis for a new production bike with decent dirt-riding capabilities. Not a lightweight single-cylinder dirt weapon like the Japanese were developing at the time, but rather a sort of "street scrambler," with the Boxer engine and shaft drive but radically reduced weight (just 186kg wet), long-travel suspension, dirt bike wheels, and a new Monolever single-sided swingarm. The new bike was called the R80 G/S, with the G/S meaning "Gelände/Strasse" (off-road/road). Those two letters (minus the typical postwar Munich slash) are perhaps the most important in BMW's recent history. They are the suffix to an entire range of GS machines, from the more recent single-cylinder G310 and F650 models, through the parallel-twin F650 to F850 models and a slew of Boxer-powered R-series bikes.

The Boxer Rebellion

Apart from the G/S, the rest of the Boxer range faced an uncertain future from the company's point of view. In theory, the old air-cooled flat-twins would taper off as the water-cooled "flying brick" K-series bikes took over in 1,000cc and 750cc form. There was an announcement that smaller Boxers would continue in production, which seemed a bit strange, since the low-capacity R-series bikes were probably the least effective. A 1,000cc twin motor that only uses a 450cc or 650cc capacity is destined to lead to a heavy, underpowered design.

Luckily, the punters were having none of it, and "popular demand" brought the old big-bore Boxers back. The R80 was relaunched in 1984 alongside the R80 RT, and the R100 RS and R100 RT followed in 1986 and 1987, respectively. All were overhauled, including the Monolever rear suspension from the G/S design, front forks and disc brakes from the K series, and a host of other updates. The engine was smoother, quieter, and more refined, but there was no power increase. The R100s had their output cut to 60 bhp, down 10 bhp from the late-1980s R100 models, and the R80 made just 50 bhp.

If even that was too much, the R65 was relaunched in 1985 with a 649cc Boxer in a basic naked roadster Monolever chassis. It made 48 bhp and was also available in a special 27-bhp version for novice riders in the German home market.

Back to the Boxers

As the 1980s drew to a close, Germany's future was suddenly rather uncertain. The Berlin Wall collapsed in 1989, and with the fall of the Soviet Union, East and West Germany seemed to be heading for reunification.

At BMW, the future of its K-series machines was also looking a little uncertain. The K1 had earned critical acclaim for its advanced, futuristic design, albeit tempered by the 100-bhp engine's unambitious power output. But as is often the case, approval from the critics didn't dictate mainstream sales success. The K1 was pricey, £7,800 new in 1989 in the UK, around 50 percent more expensive than a full-on high-octane superbike like the £5,449 137-bhp Kawasaki ZX10.

1987–1990 R100 GS

Alongside the all-new R80 GS (no slash) for 1987 came the R100 GS: a 980cc, 60-bhp Boxer engine with the latest internal upgrades, matched to the new Paralever chassis, and featuring tubeless wire-spoked wheels for the first time. It looked the part for sure, featuring a 21-inch front wheel with dirt-style mudguards and long-travel suspension, a high-level exhaust system, small headlamp fairing, and a large 24-liter fuel tank. All-up weight was 210kg, ready to ride.

The larger engine wasn't much of an advantage off-road, and the extra 10 bhp over the R80 GS was moot on most dirt roads. Where it did help was on the tarmac; the new R100 GS was a great all-around road bike. It was more than up to a forest track or fire road, and the off-road gods could easily take it around the toughest enduro courses. But for the majority of owners, it was the two-wheeled Land Rover that BMW suggested it could be, the true ancestor of today's heavyweight adventure-touring machinery.

There was another GS for 1987, this time a full 980cc R100 GS, with the new Paralever back end. Styling and kit upgrades plus the extra grunt from the bigger engine meant it worked well on road as well as off.

PARIS
DAKAR

Change was on the horizon. BMW had a larger 1100 version of the K-series powerplant ready to go into production, but it would very much be more of the same: the heavy laid-down powerplant in a series of large touring-focused bikes. Meanwhile, the Boxers really were at the end of the line, despite the resurgence provided by the excellent GS models. The basic engine design that traced its roots all the way back to the 1969 R75/5 didn't have the power or the modern design to take on the next decade. It used carbs, pushrods, two-valve heads, and air cooling in a time when fuel injection, double overhead camshafts, multivalve heads, and liquid cooling were becoming the norm.

BMW needed to come up with something new. And it did, in spades, with an all-new Boxer engine for the 1990s that would see off the laid-down K-series bikes and take the firm well into the next millennium.

ABOVE: **Could this picture be any more 1980s? The fabulous K1, some fabulous eighties big hair, and fabulous BMW riding kit and helmet.**

OPPOSITE: **A special Paris–Dakar version of the R100 GS appeared in 1989, celebrating the company's success in the famous desert race. It featured aggressive styling, engine bars, and an external headlight protector.**

EFI

German engineers—indeed, BMW engineers—had pioneered gasoline fuel injection in the first part of the twentieth century with their work on high-performance supercharged aircraft engines. Injecting or spraying fuel under pressure into an engine has a number of advantages over carburetors, particularly for forced induction, or the sort of high-energy aerobatics needed in dogfights. Basic carburetors don't work when upside down, for example, and even the early mechanical fuel-injection systems produced more power with higher efficiency. The BMW 801 radial used a direct petrol injection system and analog computer control, which was a big part of its superiority over older designs.

For motorcycles, however, which are seldom upside down (at least not intentionally), carburetors were an excellent fueling solution. In terms of peak power production, drivability, cost, and reliability, a properly set-up carburetor was hard to beat right up to the mid-2000s. Where they don't work well is in terms of engine emissions. Without fine control over how much fuel goes into the engine, the exhaust gas coming out will contain much more carbon monoxide, unburnt hydrocarbons, and other pollutants. In addition, the catalytic converters used to clean up exhaust emissions even further need to be used with an accurate electronic fuel-injection system. Laws regarding engine emissions appeared in California in the late 1960s and soon became widespread for cars, meaning automotive fuel injection was ubiquitous there by the late 1990s. Emissions rules for bikes came much later, and most models stuck with carbs until the 2000s.

BMW had dabbled with mechanical fuel injection on its RS54 racebike in the 1950s at some race meetings, but it wasn't until 1983 that it fitted fuel injection to a road bike—the new K100. It wasn't needed for emissions (yet), but BMW incorporated it as part of the high-tech overhaul that came with the K-series laid-down inline-four engine design. The first system used was a variant on the analog Bosch L-Jetronic setup used on BMW's cars, the same as that used on the first production fuel-injected bike, Kawasaki's 1980 Z1000 H. Later, in 1988, the K100 range moved to the Bosch Motronic digital fuel-injection system, with the more modern type of ECU that uses a digital microprocessor rather than analog signal-processing electronics. And in 1993, the Boxer range finally dropped carburetors, with the new four-valve Oilhead Boxer engine using Bosch Motronic digital fuel injection.

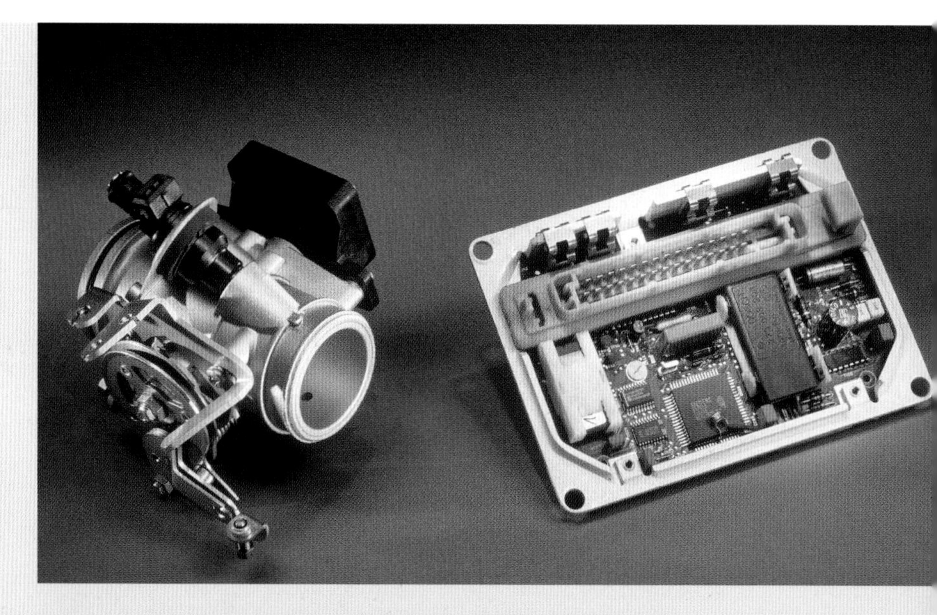

ABOVE: The 1993 R1100 RS used the Bosch Motronic MA 2.2 fuel-injection system shown here: a conventional throttle body with cable-operated butterfly valve and a digital electronic control unit.

OPPOSITE: The K1600 uses a single ride-by-wire throttle valve leading to the six-branched inlet manifold.

BMW's fuel injection worked well but remained a bit of a niche technology due to little real demand from riders. Indeed, many were scared by the notion of a computer-controlled fueling setup that had no simple way of maintaining or repairing it without main dealer equipment. Firms like Ducati and Moto Guzzi dabbled with injection through the 1990s, but it wasn't until the early 2000s that it became a mainstream technology, appearing on high-volume Japanese machinery as they met increasingly strict European emissions rules.

The next big step in fuel injection technology was the move to "ride-by-wire" systems pioneered in the car world. Here there's no direct mechanical connection from the throttle twistgrip to the engine intake; rather, the twistgrip sends an electronic signal to the fuel-injection ECU, which interprets that as a power "demand" from the rider. The ECU then uses an electric stepper motor to open the physical butterfly air valves in the engine intake while adjusting the amounts of fuel injected alongside the airflow.

Ride-by-wire reduces emissions even further by giving the ECU total control of both air and fuel flow into the engine so they are always optimized for engine revs, road speed, gear position, coolant, and air temperatures. It allows for advanced rider aids such as traction control, wheelie control, and launch control on high-performance machines like the S1000 RR superbike. It also allows engineers to easily incorporate a precise cruise control system.

1985–1996 K75 C, 1985–1995 K75 S, 1989–1996 K75 RT, and 1986–1996 K75

BMW wanted to expand the K-series range and chose a slightly unconventional way to do it. The K75 C, launched in 1985, took the four-cylinder laid-down engine design of the K100 and simply removed one cylinder to make a 740cc triple. The crankshaft was changed to a 120-degree design, and balance weights were added to a secondary shaft to reduce vibration. The bore and stroke stayed the same as on the K100 (67x70mm), but compression was raised to give the K75 a slightly higher state of tune. As a result, the three-cylinder motor made 75 bhp at 8,500 rpm compared with the 90 bhp of the K100.

The rest of the bike was largely identical to the K100: a steel-tube bridge-type frame, Monolever shaft drive, five-speed gearbox, Bosch fuel injection, and dual front brake discs. The rear wheel's 18-inch rim with a drum brake varied from the 17-inch disc-braked unit used on the rest of the K range. A small headlamp fairing gave some wind protection at speed, and the K75 C could hit a respectable 125 mph, not far off the 132 mph of the K100.

The K75 range expanded over the 1980s with the half-faired K75 S, full-faired K75 RT, and naked K75 all offering varying takes on the 750 triple setup. Some K-series aficionados actually rated the lighter, nimbler triples above the K100 fours, with the small power loss more than made up for by sharper handling.

ABOVE: The K75 S launched in 1985, with the new three-cylinder 750 version of the K engine in a sporty half-faired chassis.

RIGHT: A pre-production version of the K1 on a handling test circuit. Note the super-slick aerodynamic front fender and massive Paralever rear swinging arm.

1988–1993 K1

In many ways, it was a bit of a surprise that the original K-series engine didn't have a sixteen-valve cylinder head. Many air-cooled bike engines in the 1970s had four valves per cylinder, and the design was starting to appear on the latest generation of water-cooled sportsbike engines from Japan. A sixteen-valve head would have gone beautifully with the double overhead camshafts, fuel injection, and water cooling on the first K100 engine.

The new sixteen-valve K100 engine finally appeared in 1988, but rather than appearing first in the standard K100 range, BMW wrapped the new engine in a wild new fully faired superbike: the K1. With a name that echoed the firm's M1, M3, and M5 supercars, the K1 was an audacious assault on the Japanese stranglehold at the top of the bike world.

The redesigned sixteen-valve DOHC cylinder head was matched to higher-compression pistons, a new digital fuel-injection system, lightened rods and crankshaft, and tweaked gearbox. The bore and stroke stayed the same at 67x70mm, while the power output now touched 100 bhp, up 10 bhp from the original K100, in line with the voluntary German motorcycle power limits of the day. The powertrain was fitted with the new Paralever rear suspension off the latest GS models, and the main frame was modified with larger-bore steel tubing for extra stiffness. Four-piston Brembo brake calipers up front provided class-leading stopping power (ABS was optional in most markets), and suspension was by high-end Marzocchi forks and a Bilstein rear shock.

The most obvious element of the K1 is the bodywork. Designed with wind tunnel research to reduce aerodynamic drag to the absolute minimum, it was like nothing else ever seen. From the all-enclosing front mudguard back to the angular fairing lowers and massive tail unit, the aim was to guide air smoothly with minimum drag around bike and rider. Loud, aggressive paint schemes—red/yellow and dark blue/yellow with bright yellow wheels and swingarm—underlined the radical nature of the latest BMW. We were a very, very long way from the days of the original all-black BMW Boxers.

The sixteen-valve K1 engine appeared in an updated K100 RS in 1989, alongside the rest of the K1 upgrades. Indeed, the sixteen-valve K100 RS was basically a K1 in traditional Bavarian clothing, giving more traditional fans a less exotic option.

The K1 chassis design could have been even more radical, it seems. Hidden in the BMW archive is this picture of a prototype K1 front wheel with a Buell-style rim-mounted brake disc.

This poster of a cutaway K100 RS sixteen-valve would have graced the bedroom wall of many a two-wheeled geek in 1989. The author didn't have one but would certainly have made space for it.

BMW

ABS

Like fuel injection, the roots of anti-lock braking systems come from the aviation world. The job of braking a heavyweight jet aircraft after it lands on a runway is not trivial—braking too hard could lead to a tire blowout and crash, but not braking hard enough can see a plane crash off the end of a runway. On a more prosaic level, inefficient braking wears out expensive plane tires fast, costing companies a fortune.

There was a big push for anti-lock braking from plane manufacturers, and the first system was produced in the 1950s by British firm Dunlop. It was a mechanical setup and was too expensive and heavy for road use. But the advantages on a bike are even more obvious than on planes: lock the front wheel of a motorbike unexpectedly at speed under braking and you will have a crash nine times out of ten, with all the risk that entails.

Fast-forward to 1988, then, and BMW was unveiling the first production motorcycle ABS system on its updated K100 range. By now, ABS was common on high-end cars, and the K100 used a modified version of the system on BMW's four-wheeled range, which was developed alongside the FAG Kugelfischer firm.

The principle is simple enough—toothed ring sensors measure the rotational speed of each wheel. An ECU compares both speeds with preset figures for wheel deceleration. If one wheel keeps turning and the other stops, or if a wheel is slowing down much faster than it could ever do under the hardest of braking, then the ECU decides that that wheel must be locked, or is about to lock up. It then sends an electrical signal to a solenoid valve and hydraulic pump mechanism to momentarily release the brake pressure to that wheel and then reapply it a fraction of a second later.

Early systems were heavy (7.6kg for the two Kugelfischer hydraulic pump units alone), were fairly basic, took a rather long time to reapply the brakes, and were ineffective during cornering. As technology improved, the hydraulic units cycled on and off much faster, and more powerful ECU components could process wheel movements much more accurately.

BMW has remained at the forefront of ABS technology, and its latest systems use an IMU inertial measurement unit in the ECU, so the bike knows if it's cornering or going straight and whether the front or back wheel are off the ground. The system can provide maximum braking effort without locking in a bend and prevent the rear wheel from lifting off the ground when hard on the front brakes. The system can also incorporate other useful functions like hill start assist, where the bike can hold the rear brake on at a standstill until the rider releases the clutch to pull away.

ABOVE: This schematic shows BMW's Race ABS setup on the 2015 S1000 RR. A compact ECU/pump unit controls the pressure in both hydraulic circuits and has advanced anti-locking algorithms tuned for circuit use.

LEFT: This K100 test mule was used to develop BMW's early ABS prototype systems. Computerized test kit was, of course, much bigger than today, hence the huge electronic rig mounted above the dash. The basic principles of this test rig are the same as on current ABS setups, however: wheel-speed sensors, electronic hydraulic pumps, and a computer algorithm watching for locked wheels, then reducing brake pressure to avoid it.

ABOVE: Braking hard on slippery surfaces has little margin for error on a bike. What would be a minor skid on a car would often end up in serious injuries or fatalities before ABS became common on bikes. BMW was the main pioneer in this technology, with production ABS systems available long before the competition.

LEFT: BMW's early production ABS setup, as seen here on the K100 LT, was heavy and intrusive in operation. It couldn't cope with corners, would cut in when braking over bumps, and was generally a bit primitive. But it was still a lifesaver for many, and BMW soon improved the performance massively.

K1100RS

7

1990s: Defining Performance

At the beginning of the 1990s, the world was moving fast. West and East Germany reunited with minimal fuss (just as West Germany won the 1990 soccer World Cup), the Soviet Union was finally gone, and war was brewing in the Persian Gulf and in the Balkans. The internet was born, email was spreading out of university computer departments, and mobile phones shrank from the size of a brick to the size of a jacket pocket.

Motorbikes were moving fast too, literally and metaphorically. Kawasaki launched its 175-mph ZZ-R1100 in the spring of 1990, and it immediately topped sales charts across Europe. Even more of a threat to BMW was the Honda ST1100 Pan European, a shaft-drive, 1,100cc heavyweight tourer, with strong performance and the superb build quality that defined Honda motorcycles from the early 1990s.

The Japanese dominance of the previous decade wasn't going away anytime soon, and there was more competition coming from Europe too. Ducati was resurgent in the high-performance superbike market, while in England, businessman John Bloor was resurrecting the defunct Triumph brand (with a little help from Japan, ironically).

Bigger, Better?

BMW wasn't hanging about either. The K-series bikes got another upgrade, this time taking the capacity to 1,092cc on the K1100 LT luxury tourer and K1100 RS in 1991 and 1992. Both the K1100 LT and RS used the same K1-based chassis setup, but the RS had a lighter, sportier package. The K1100 models also came with the next generation of BMW's ABS system as an option. The new ABS II components were much lighter, and performance was improved all around.

Despite two cracking big bikes, the K series was arguably heading in the wrong direction in some ways. The point of a four-cylinder engine is to give more top-end power than a twin, but the K1100 units were tuned for more torque, not more power, and kept the same 100-bhp peak output as the sixteen-valve K1 engine. Why not, then, go back to a twin-cylinder motor? It would have less weight and complexity than the sixteen-valve inline-four unit, a little less power, and the strong, torquey power delivery you were tuning the K engine to make. A new Boxer maybe?

The handsome K1100 RS performs the difficult trick of remaining true to the style of its ancestors while bringing it right up-to-date.

ABOVE: **The original laid-down K series still had plenty of life left in it. The K1100 series used a big-bore version of the sixteen-valve K100 engine, in both this sporty RS variant and the full-dress LT tourer. Both would be developed into 1200 versions that lasted well into the twenty-first century.**

RIGHT: **The huge top box and radio antenna mark the K1100 LT out as a genuine luxury tourer. Note the deep padding on the pillion seat and back rest: there are many tales of sleepy passengers nodding off on the back of a continent-crossing LT.**

1991–1999 K1100 LT

By 1991, the days of motorcycle touring as invigorating but spartan transport were long gone, thanks mostly to firms like BMW. Touring bikes like its K100 LT and Honda's Gold Wing came straight from the factory ready to cross continents in comfort with rider, passenger, and luggage.

The K1100 LT took the premium touring sector to a new level thanks to massive protective fairing, sofa-style seating for two, and extensive touring accessories—hard luggage, a radio cassette deck, heated grips, and an electrically adjustable windshield. The LT also had the option of a fully integrated three-way catalytic converter in the exhaust to remove hydrocarbons, carbon monoxide, and nitrogen oxide pollutants, another piece of now-ubiquitous bike technology with which BMW was first out of the blocks.

It was the first application of the new K1100 sixteen-valve powertrain, with Paralever rear suspension, telescopic front forks, and the stronger steel-tube bridge frame from the K1. The 1,092cc engine capacity boost came from a larger bore (up to 70.5mm from the K100's 67mm, which was right on the block's limit. (Later K1200 models had to use even longer stroke motors to increase the cubic capacity.) Power stayed at 100 bhp but had more torque, and though the 1100 LT tipped the scales at 290kg ready-to-ride, it handled well. The laid-down K motor's low center of gravity made the weight more manageable at low speeds, and once moving, the Paralever rear suspension and 41mm front forks provided a comfortable but well-damped ride that was much more dynamic than the super-heavyweight Gold Wing. Later on, special Highline and SE editions added a host of optional accessories as standard.

ABOVE: Extensive instrumentation, adjustable windshield, power socket, vanity lighting—the LT was well ahead of the pack in terms of luxury touring kit.

LEFT: In the 1990s, the Highline special edition of the K1100 LT was hard to beat as an overall luxury touring package. The Honda Gold Wing was a strong contender, but the LT was arguably the better machine on European roads, while the Honda ruled on American highway routes.

1992–1997 K1100 RS

If the K1100 LT was just too big for you, the K1100 RS offered a slightly less gargantuan sport-touring take. This was BMW sport-touring, mind, rather than the more lightweight concept used with Honda's contemporary VFR750 (which was around 50kg lighter with the same peak power). The K1100 RS tipped the scales at 268kg ready-to-ride, thanks to more compact bodywork based on the K100 RS with a lower fairing and dynamic engine air venting. It had less kit than the LT but stuck with the same engine and chassis package, including the ABS and catalyst options. It had a certain cachet and could go up against the likes of Kawasaki's ZZ-R1100 in terms of handling comfort and mile-munching. It was well down on outright power, however, making just 100 bhp compared with 147 bhp for the Japanese machine.

The Oilhead Boxer

There had been rumors of a new Boxer engine all through the late 1980s, hardly surprising when you consider both the popularity of the big twins and the rather slack progress of their development at that point. Granted, the K-series bikes were meant to take over the BMW range, but to still be selling the R100 RT in 1996 whose engine was designed in 1969 is pretty shocking in retrospect.

In 1993, the big flat-twin motor from Bavaria took another one of its quantum leaps forward with the launch of the R1100 RS. Absolutely everything about the bike was new, from the four-valve fuel-injected engine to the Telelever front suspension, frameless chassis, and all-enclosed full fairing. Looking back at it now, it screams "1990s" like an early episode of *Friends*, but back in 1993, it was as futuristic as an F-117 stealth bomber. In comparison, the R100 RS it replaced looked more like a World War II–era B-17 bomber.

The R1100 RS was merely the vanguard for the new Boxer powerplant and its associated chassis package. The R1100 GS version followed in 1994, alongside the naked R1100 R roadster, then the heavyweight touring-oriented R1100 RT in 1995. A whole new era was about to begin for BMW's Boxers.

Meet the new boss—BMW's 1993 R1100 RS took the Boxer paradigm of the R100 RS to all-new levels, with fuel injection, oil cooling, advanced front and rear suspension systems, and ABS, but still with the fundamental shaft-drive, flat-twin ethos. It was a big moment for the Bavarian firm.

4VALVE

The R259 Boxer Oilhead Engine

Oil cooling was nothing new in the 1990s: Japanese bike company Suzuki had made it the focus of its high-end superbikes all through the 1980s. Adding a radiator for the lubrication circuit was also common on even medium-performance air-cooled engines of the time. It was a bit of a no-brainer: you already had an oil pump delivering lubricating oil all around the engine, so why not add some extra piping, a simple bypass valve, slightly larger oil capacity, and a little radiator under the steering head? Removing heat from the oil reduced temperatures around hot spots in the combustion chamber and pistons, and with a few more mods like piston cooling jets and a separate cooling circuit with its own high-volume pump, you could almost match the power output and reliability of water-cooled engines with less weight and complexity.

BMW took all these steps with the new R1100 Boxer engine launched in 1993. Together with the natural cooling advantages of the Boxer layout, the oil-cooling setup allowed a 50 percent power increase over the old 980cc R100 motor without overheating. It also gave the new engine its nickname—the Oilhead.

The Oilhead's power boost—up to 90 bhp—came in part from its larger capacity, up to 1,085cc from a 99mm bore and 70.5mm stroke. Compression was boosted to 10.7:1, and there was an all-new four-valve cylinder head design with fairly large valves: 36mm inlets and 31mm exhaust. The new 99mm cast-aluminum three-ring pistons were lighter than the old air-cooled Boxer parts and ran inside silicone-nickel-coated cylinder bores.

The design engineers couldn't use a "normal" high-performance DOHC setup on the Boxer; having large camshafts and covers on the end of the cylinders would make the engine unfeasibly wide. They adopted a clever "high cam" design that had a single chain-driven camshaft in the cylinder head on each side, mounted offset from the combustion chamber. Special short pushrods then operated rocker arms in the head to open the paired intake and exhaust valves. It was a good compromise, much narrower than a DOHC design, and the rpm limitations of a pushrod design were minimized by the compact design. The super-short, stiff rods could cope with higher engine revs than the old Boxer's full-length pushrods, and the new engine made peak power at 7,250 rpm.

The bottom end was just as innovative, with a plain-bearing crankshaft, vertical-split crankcases, and forged sintered steel con rods that used the then-novel fracture-split big end caps. Carefully "breaking" the con rod at the big end cap faces gave a joint with much more surface area than even the finest cutting saw, increasing strength across the joint with no weight penalty. There were just two main bearings, with no bearing between the two big end journals, making the engine shorter front to back. The alternator was mounted above the crankshaft, fitting into the space inside the front subframe, and was driven by a rubber V-belt.

Fuel injection was well-established technology for BMW by now, and the R259 used the latest variant on the Bosch Motronic MA 2.2 digital engine management system. That also allowed a three-way catalytic converter as an optional factory-fit item.

Another vital part of the new Boxer engine was actually its chassis function. BMW decided to use the massive new crankcases as the main part of a new frameless design. The new Telelever front suspension arm pivoted on the crankcases, right above the cylinders, and the single-sided rear Paralever swingarm was already an integral part of the powertrain. Two simple steel-tube subframes above the engine mounted the steering head at the front and the rear suspension top mount, seat, and tail unit out back.

OPPOSITE: With its ABS-equipped Brembo brakes, sleek full fairing, fuel-injected four-valve motor, and single-sided rear swingarm, the R1100 RS was a high-tech beastie in the early 1990s.

ABOVE: This rear-view cutaway drawing of the R1100 RS motor shows the locations of the camshafts high up in the head as well as the cam chain drive setup.

1992–2001 R1100 RS

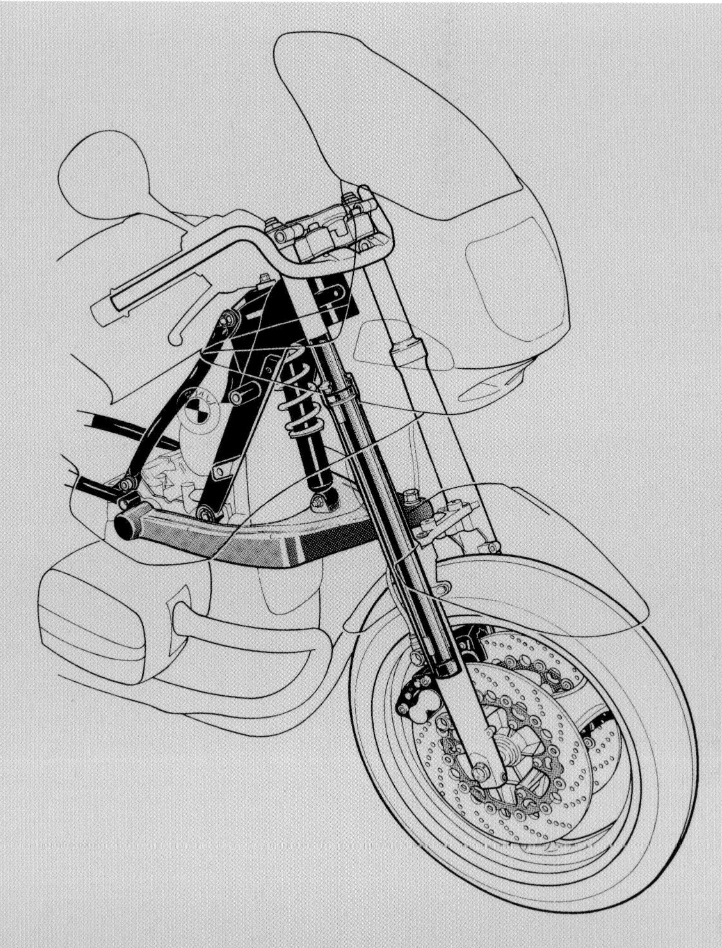

The R100 RS had revolutionized BMW's big twins in 1976 when it launched, and sixteen years later, the R1100 RS took the sporty touring RS into the next generation. It was the standard bearer for the new Boxer engine featuring four-valve heads, high-mounted camshafts, and fuel injection, as well as the new Telelever front end and frameless chassis. It was a bold move from Munich—and a risky one. Going for a total revolution in both engine and chassis design at the same time is a huge undertaking and can make development much harder.

But BMW's engineers had cracked it, and the R1100 RS was a huge performer right off the bat. The new engine made 90 bhp, more than enough for a sports tourer at the time, especially with the torque from a big-twin layout. The new transmission design was BMW's typical dry car-type clutch and five-speed gearbox, missing the chance to move to a six-speed unit.

The new chassis impressed with its handling (the Telelever front end worked well on this type of bike), and it came with wide, sporty radial tires, a 120/70 17 front and 160/60 18 rear. The front brake used the high-end Brembo four-piston calipers first seen on the K1 and dual 305mm discs. The latest ABS setup was an option. Seat, handlebars, and windshield were adjustable to fit to individual riders, and BMW's accessory catalog offered a wide range of hard luggage, heated grips, crash bars, and other useful add-ons.

TOP: This early design sketch shows how the Telelever front end was such an integral part of the design. The engine was designed to work with the chassis and vice versa. Neither would have been possible in isolation.

RIGHT: This shot of the R1100 RS engine shows how compact the powertrain is. There's a lot crammed in there: note the large mount points for the Telelever pivot arm above the cylinders, the EFI throttle bodies, and the alternator positioned above the front of the crankcases.

OPPOSITE: Sporty touring models were huge in the European market back in the 1990s, and the R1100 RS made an instant impact.

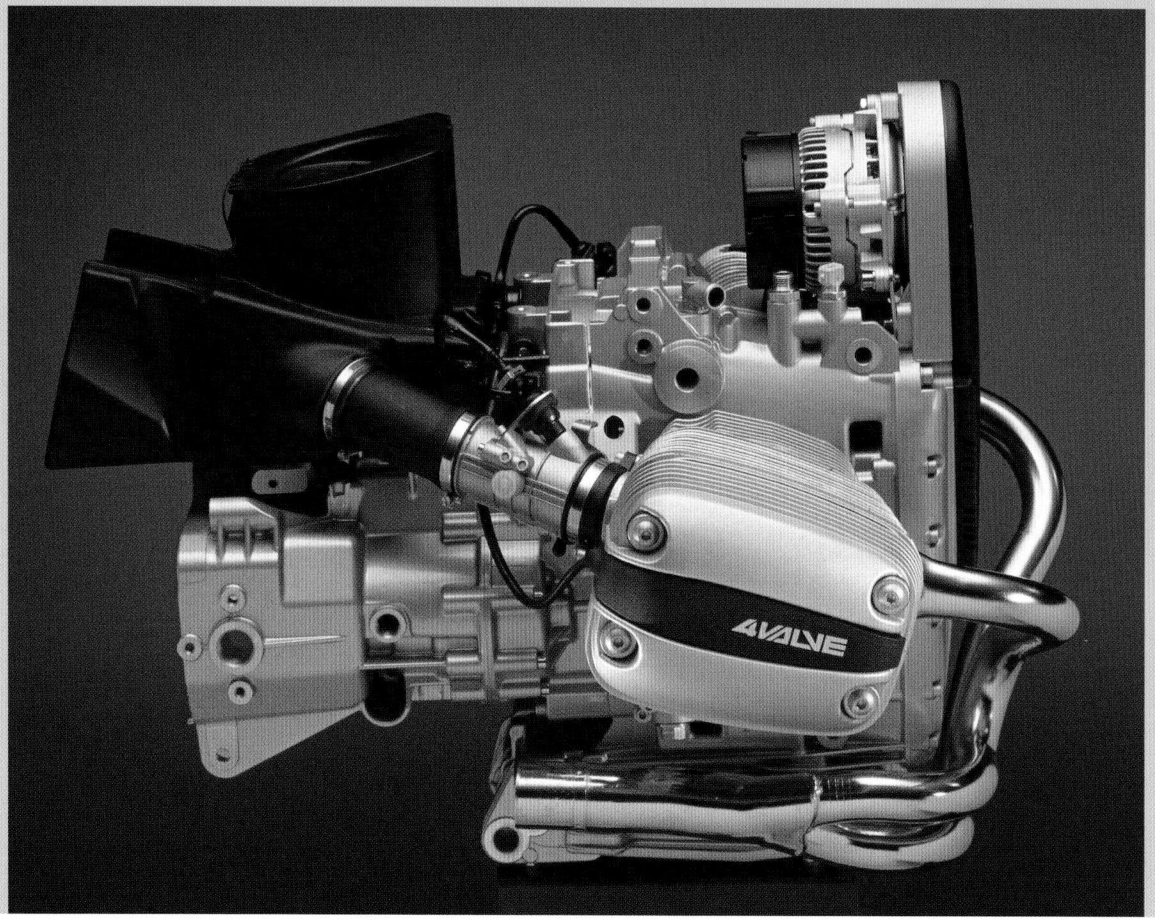

BMW
4VALVE

Telelever Front Suspension

The traditional front fork used on the vast majority of motorcycles is a bit of an engineering nightmare. Having the entire set of front wheel forces—suspension, steering, braking, and leaning—transmitted to the frame by a pair of long, sliding metal tubes with springs inside brings a whole slew of compromises. It's a testament to human ingenuity that modern fork designs work so well.

Plenty of designers have come up with alternatives over the past hundred years, and BMW, for better or worse, has been an enthusiastic adopter of odd front ends. Its pioneer bikes like the R32 used a simple leaf spring trailing link front suspension before a move to "proper" hydraulically damped telescopic forks for the 1935 R12 and R17 750s. But in the 1960s, the firm opted to fit the then-novel Earles front end, which had a swinging arm and dual shocks. The Earles design was great for sidecar use, but it was far too heavy for high-performance solo machines, and by the early 1970s, BMW went back to telescopic forks.

It was a grudging compliance with the norm, though, and the next experiment came in 1993 with the R1100 RS Boxer. BMW had decided to adopt a new setup: a swinging arm mounted on the engine crankcases and linked to a pair of sliding telescopic legs that looked like forks but had no springs or dampers inside. The suspension function was performed by a single monoshock unit between the swinging arm and the front subframe. A spherical bearing linked the arm to the sliders, allowing normal steering movement all through the suspension travel.

The Telelever setup optimized braking forces, giving a direct structural link from the front wheel through the lower sliders and into the swinging arm back to the engine, which let the designers tune in the amount of dive they wanted when braking. It was also stiffer, since the front wheel forces only traveled as far as the beefy swinging arm rather than all the way up to the steering head.

The Telelever was (and remains) a good solution for BMW's bigger machines. It's particularly well suited to the full-sized GS models, where it allows plenty of front suspension travel without excessive movement under acceleration and braking, and adds strength and stiffness for off-road antics. The R1250 GS Adventure would be a very different machine without its Telelever front end.

On the other hand, the system isn't so good for smaller or cheaper bikes, or high-performance sportsbikes. It's more expensive to manufacture than a simple set of forks, and it's a heavy setup overall, so it doesn't feature on the F and G series of singles and parallel twin machines. Add on the difficulty in packaging the Telelever swinging arm mounts onto a short-wheelbase four-cylinder engine, and it's obvious why the firm's S1000 RR, S1000 R, and S1000 XR all use conventional (if high-performance) telescopic forks.

This later version of the R1200 GS shows another angle of the swinging arm and the spherical joints at the top of each fork leg. These allow for the movement of the fork top mounts as the long-travel suspension moves up and down.

Middleweight Options

BMW's rather meager middleweight offerings received a boost in 1995, when the R850 R was released. It used a sleeved-down version of the 1100 Boxer motor and an 87.8mm bore instead of 99mm, cutting capacity from 1,085cc to 848cc. Power dropped to 70 bhp, but since the rest of the bike was basically the same as the R1100 R, the 235kg (wet) weight was just too high for that power. Over the next few years, an R850 GS and R850 C cruiser appeared with the same formula, but none of them were very popular. There were simply too many better, cheaper, more dynamic options in the middleweight market.

The last of the 1969 Boxer's direct descendants was released in 1996, the 980cc air-cooled R100 R Mystic. This model was aimed at the hard-core BMW believers for whom the oil-cooled, fuel-injected Boxer was just too much. There was also an R80 GS Basic, celebrating the carbureted, air-cooled, two-valve twins as the century drew to a close.

BMW also offered a basic naked roadster, the R1100 R, which was also available in (virtually) visibly identical 850 form shown here. The R was far from the best of the new Boxers, but it sold well enough in the home market.

BMW
brembo

The Return of the BMW Single

BMW had dumped its rudimentary single-cylinder machines when the archaic R27 was discontinued in 1966. In the following twenty-five years, the firm had fairly poor entry-level offerings, such as sleeved-down 450 and 650 Boxer motors bolted into low-spec chassis. The likes of the R45 and R65 were heavy and underpowered and held little appeal for fans of "proper" large-capacity BMWs or riders looking for smaller machines with decent performance. Through the 1970s and 1980s, the big four Japanese companies could offer 250, 350, 400, and 500 class machines with two- or four-stroke engines, excellent power outputs, and sharp handling at affordable prices.

BMW was missing out on a number of fronts. It wasn't serving the large market for smaller-capacity machines, but perhaps more important, it was losing the next generation of riders. If your first bike is a Honda bought from a Honda dealership, there's a good chance your second and third bikes will be Hondas too. Getting novice riders to buy into a brand is an important part of most bike firms' marketing plans.

The launch of the new BMW F650 Funduro in 1993 was a big step for the company. BMW dealers could once again offer newly qualified riders a lightweight machine with modest power and nimble handling at a good price. It was a very unconventional machine for BMW. The bike itself was actually a version of the Aprilia Pegaso, powered by a Rotax single-cylinder engine made in Austria. The F650 was assembled at Aprilia's Noale factory in Italy, and some Bavarian traditionalists clutched their pearls at the thought of this Venetian usurper wearing the blue and white BMW roundel on its tank.

In fact, they had nothing to worry about. BMW kept a close eye on the quality processes, and Aprilia's production lines lived up to Spandau's standards. The Funduro's look was styled by one of BMW's own design staff (a Scot named Martin Longmore) and had a number of fundamental differences from the Pegaso, especially inside the engine. There was a congruence of style between the new single and the new R1100 RS Boxer too, and it was a big part of BMW's 1990s design feel.

It's hard to think of a starker contrast than comparing BMW's previous single with the F650 Funduro. The last mono-cylinder machine was the prehistoric R27 from 1966, a bike that even its biggest fans would admit offered little in the way of fun. The F650 was built by Italians, with German oversight, and added massively to the appeal of the BMW range in the 1990s.

BMW kept up the tradition of offering smaller-capacity versions of the big Boxers with the R850 GS. It was virtually identical to the 1100, and as this image shows, it was almost impossible to tell them apart at a glance, graphics aside.

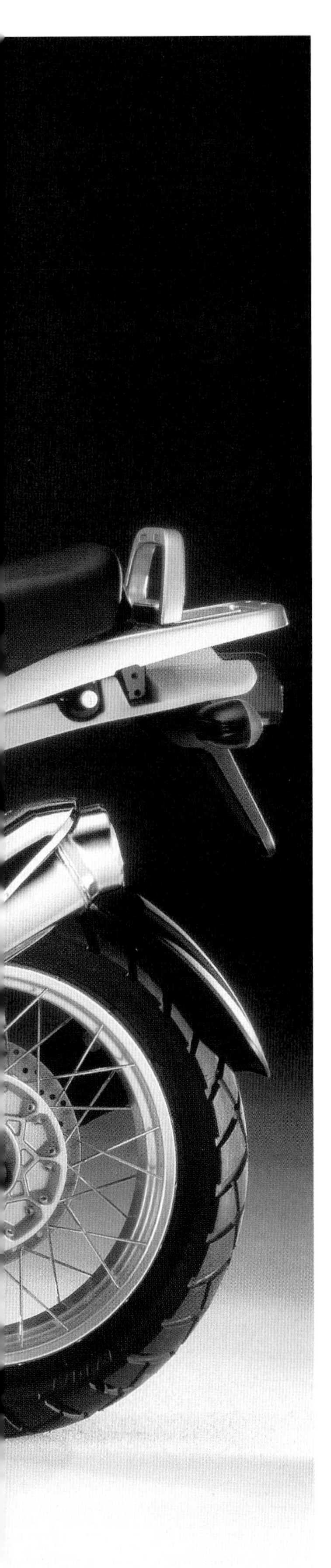

1993–1999 R1100 GS

This is arguably the bike that transformed BMW's fortunes while cementing the GS range's place in history. Before the R1100 GS, the R100 and R80 versions had been good bikes with solid performance, but they were nowhere near mainstream. While their modest power outputs, quirky chassis, and rally styling had secured niche success, the idea that they could one day top the sales charts in major markets across Europe seemed way off.

The Oilhead GS changed all that, thanks to a number of factors. First, it had (just) enough power to be exciting on the road: 80 bhp is a fair number for the tarmac, and the R1100 GS was able to stay in touch with its rivals. Second, the Telelever front end transformed the handling. Telescopic forks on a long-travel dirt bike weighing nearly 250kg would have been a wobbly, soggy disaster, see-sawing into and out of a corner. The Telelever front end allowed the long travel needed for off-roading but also permitted good wheel control and rock-solid handling. I remember in 1998 working at *Motorcycle News* and seeing photos of road tester Chris Moss riding his long-term test GS hard around a country back road. The sight of a 300kg bike/rider combo, with a kneeslider skimming the deck, relying only on a skinny 110-section front to keep things shiny-side-up has stayed with me ever since. This was a bike with far better performance than you could ever expect, and it simply worked amazingly well.

Compared with the original R1100 RS, the GS engine was detuned by 10 bhp at the top end, with slightly lower compression and better low-down performance for off-road use. The suspension units were modified for longer travel, the dirt-sized wheels and tires were fitted, and the bodywork was completely revised, including a large "beak" front fender, a high-mount silencer, an aluminum rear rack, and a 25-liter plastic fuel tank.

The later R1150 GS would build on the 1100's success, and in the twenty-first century, the 1200 and 1250 versions would take the GS to another level. But without the massive leap of the 1100, they would have had much more work to do.

LEFT: GS had become an integral part of the BMW brand, and the R1100 version was a massive leap forward. But when it was launched in 1994, no one could have predicted just how big an impact it would have on the BMW Motorrad.

FOLLOWING PAGE: Motorcycling has always been about getting closer to the scenery. With a GS, you can travel thousands of miles in comfort, then get right inside the landscape at the end of the trip.

1993–1999 F650

The F650 Funduro was arguably the most conventional motorcycle to wear the BMW logo when it was launched in 1993. Based on the Aprilia Pegaso 650 dual-sport machine, it was powered by a single-cylinder 652cc water-cooled, four-valve engine, with a DOHC cylinder head, two spark plugs, twin Mikuni carburetors, and dual exhaust headers, making 48 bhp at 6,500 rpm. The Rotax engine had an oversquare cylinder layout, a bore of 100mm and stroke of 83mm, a 9.7:1 compression ratio, a gear-driven balance shaft, dry-sump lubrication, and a five-speed gearbox. Most radical of all was the final drive: the F650 was the first BMW bike to use a chain rather than a shaft to transmit power to the rear wheel.

The chassis was unremarkable soft-enduro fare: a steel-tube cradle-type frame, conventional front forks, and an aluminum monoshock swingarm, plus wire-spoked wheels and mild dirt tires, 19 inches in front and 17 inches at the rear. Single brake discs front and rear looked after stopping duties, while a small, half-fairing, 17.5-liter fuel tank, high-level exhaust, and wide handlebars rounded off the design.

The F650 was light, at just 176kg dry or 189kg ready-to-ride, and while it was no hard-core off-roader, it was more than up to gentle green lanes or fire roads. Change the tires for more dirt-biased fitments, and in the hands of a good rider, the little Funduro could take on some fairly serious terrain.

Compared with the Aprilia, the F650 had a four-valve twin-plug cylinder head rather than a five-valve single-plug design, and BMW swapped the roller-bearing crank for a plain-bearing design. The F650 also used a tougher steel frame rather than the lighter aluminum design on the Pegaso.

ABOVE: The F650 had a clean, simple dashboard layout and more than a hint of Aprilia about the design. A small BMW logo reminds you where you really are.

LEFT: The F650 was momentous in one way: it was the first BMW bike to use chain final drive instead of a shaft, sacrilege to the BMW cognoscenti at the time.

FOLLOWING PAGES: The F650 offered some soft off-road ability, on par with much of the so-called Dual-Sport competition at the time. It would hold its own against a Honda Dominator or Kawasaki KLR650 Tengai, but it was less capable than more serious dirt machinery.

The face of the man behind the strong, clear overall design ethos of BMW in the 1990s and beyond—David Robb.

Millennial Moves

As the 1990s drew to a close, BMW was in a state of flux. The F650 Funduro was a success, and a new single-cylinder model range began to develop. The F650 ST Strada version appeared, featuring a more road-biased chassis, lower seat, smaller wheels, road tires, and shorter suspension travel. The R1100 Boxer range remained strong too, with the GS building a strong reputation as it developed the new adventure touring class almost single-handedly. The RS and RT touring bikes were doing the bread-and-butter BMW thing, keeping traditional fans happy and selling in the thousands to police and other government agencies. Quirks like the R1200 C cruiser appeared, aimed at the massive American market, and the R1100 S, an attempt to add a traditional S sports model to the new Boxer range. From far left field came the C1, a 125cc scooter with a roof, incorporating a safety rollover cage that was strong enough to protect the rider inside as much as a small car. It was an attempt at urban mobility innovation. While it would perhaps be applauded today, in 1999 it was an oddity in the mold of the Sinclair C5.

1994–2001 R1100 RT

The R1100 RS was the sporty touring Boxer, and for 1995, BMW added a heavier option in the form of the R1100 RT. The powertrain was largely unchanged, but the chassis was tweaked to handle the extra mass (it was the heaviest Boxer to date at 282kg wet). ABS came standard. The enormous whale-like enclosed fairing was supremely effective at protecting the rider and passenger, at the cost of mass and size. Optional-fit accessories included a cavernous top box to match the standard panniers, a cassette-radio with front and rear speakers, heated grips, an electrically adjustable windshield, a power socket, and much more.

If you needed a more touring-biased Boxer, the 1995 R1100 RT was just right for the job. Its larger fairing, standard panniers, and extra equipment made it a big hit with distance fans. It was ideal for emergency service users too; police and paramedic RTs were a common sight on European roads in the 1990s.

The K series was beginning to look like a developmental dead end. The laid-down engine was just too compromised in too many ways, and while the K1200 LT and K1200 RS would bravely soldier on with the powerplant, the firm was beginning to think about its next move in the four-cylinder sector.

In contrast, styling was one area where BMW was clearly progressing. The arrival of an American design chief, David Robb, in 1993 saw a more coherent design language across the range. The R1100 GS and R1200 C, for example, were very different bikes, but it was clear they had the same bold styling philosophy behind them. Robb would maintain that ethos as BMW Motorrad's chief designer until 2012. The bikes wouldn't always be pretty—BMW was never going to challenge the likes of Ducati or MV Agusta in that way—but they would continue to combine motorcycle form and function in the striking, modernist, uniquely BMW fashion.

1996–2005 K1200 RS

The K1100 RS had been bravely flying the Bavarian flag in the unlimited sport-touring class since 1992 as BMW focused on the new Boxer and F650 projects. By 1997, however, it was wildly outclassed by the likes of Kawasaki's ZZ-R1100 and Honda's CBR1100 XX Super Blackbird. Minor tweaking wouldn't be enough to get the K-series RS back in the game though, and BMW went all in on a totally new design. The laid-down K-series engine fundamentals remained, but everything else went in the bin. The engine had no space left for a bigger bore, so designers fitted a new longer stroke crank (up to 75mm from 70mm), raising capacity to 1,171cc, added high-compression pistons with forged con rods, and made a host of other mods. The end result was impressive: 130 bhp, by far the most powerful BMW bike engine ever.

The chassis changes were even more radical. The old steel-tube bridge frame, with the engine as a stressed member, was dumped for a new fabricated aluminum semi-monocoque design that held the engine in new rubber mounts. The point was to reduce vibrations from the uprated engine, but with rubber mounting, the engine couldn't act as a structural frame component anymore. The new frame had to be stiff enough on its own, and the beefy construction brought a big weight increase despite aluminum construction.

The K1200 RS also replaced the 1100's forks with a variant on the latest Boxer's Telelever front suspension setup, modified for sportier handling. Brakes, wheels, and tires were all uprated too, with 170-section rear rubber marking another new record for a BMW bike.

The massive effort paid off, in a fashion. The K1200 RS was a hugely impressive machine, but was flawed by its sheer mass and size. At 285kg wet, with a 1,555mm wheelbase, it was too heavy and long to make it as a sports machine, and its touring skills were compromised by the sporty riding position and smaller fairing. Once you started adding hard luggage and a passenger, it all got even bigger and heavier, and you were left wondering just what the point was.

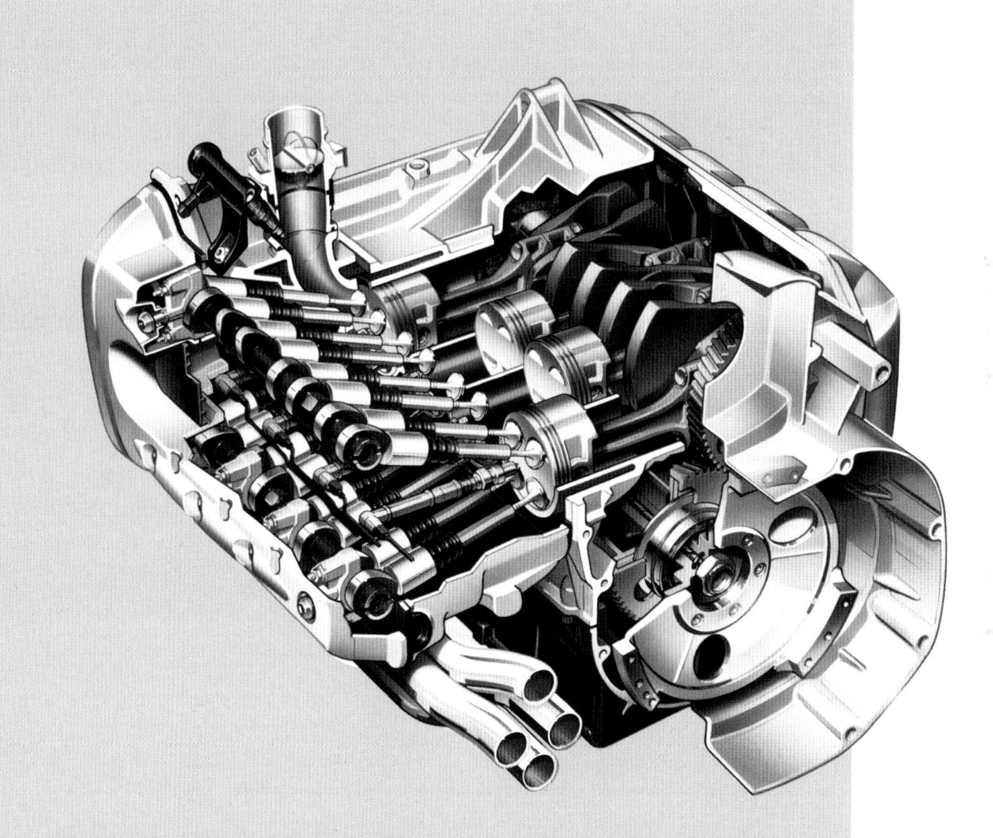

ABOVE: Producing 130 bhp from 1,170cc was fairly impressive in 1996, especially considering the K1200 RS engine's roots in the early 1980s K-series design.

FOLLOWING PAGES: The problem many riders have with a bike like the K1200 RS is that it's a bit too heavy to work as a proper sportsbike, yet it has less touring space and kit than a "proper" tourer.

BMW
BMW

1997–2005 R1200 C

Every now and then, BMW drops something so surprising that you're not sure quite what to make of it. And in 1997, it came in the form of the R1200 C, a custom version of the R1100. It had a bigger engine, apehanger handlebars, a stretched swingarm, and a kicked-out front end. The Telelever system remained, with highly polished aluminum and chrome plate everywhere, and there was even a custom-style single seat unit, complete with a sissy bar backrest that pivoted down into a small passenger pad.

The engine was the first 1200 Boxer and actually used the same bore and stroke as the later R1200 GS (101x73mm) for a 1,170cc capacity. The R1200 C unit was detuned for more midrange grunt, with smaller valves and soft cams, and produced just 61 bhp.

The R1200 C was aimed at the US market, where it had some niche sales success. Arguably its biggest impact was in the 1997 James Bond movie *Tomorrow Never Dies,* where it featured in one of Bond's trademark stunt-studded car chases.

This Independent (called the Phoenix in the US) was one of several factory-custom variants of the R1200 C, with a single seat, accessory lights and flyscreen, and premium cast wheels.

1997–2005 R1100 S

The R1100 S was BMW's first attempt at a modern sporting Boxer. A drastic overhaul to the engine and chassis helped it begin to approach the power and weight levels of 1990s machinery. The basic R1100 Boxer powertrain was tuned to make just under 100 bhp, thanks to high-compression pistons and stronger con rods allowing more revs and more power. A swooping exhaust system led to an underseat dual silencer setup, and there was a six-speed gearbox to suit the slightly peakier power curve.

The chassis used a curious new aluminum main frame that linked the swingarm pivot area with the front Telelever pivot point. That boosted stiffness over the R1100 RS, which just used the engine as the structural member but also added weight. And the 229kg curb weight, together with the shaft drive, put limits on the R1100 S's ultimate track performance. Despite that, BMW promoted a one-make Boxer Cup race series from 1999, which pitted retired world-level racers against keen club racers in support races at Grands Prix race meetings.

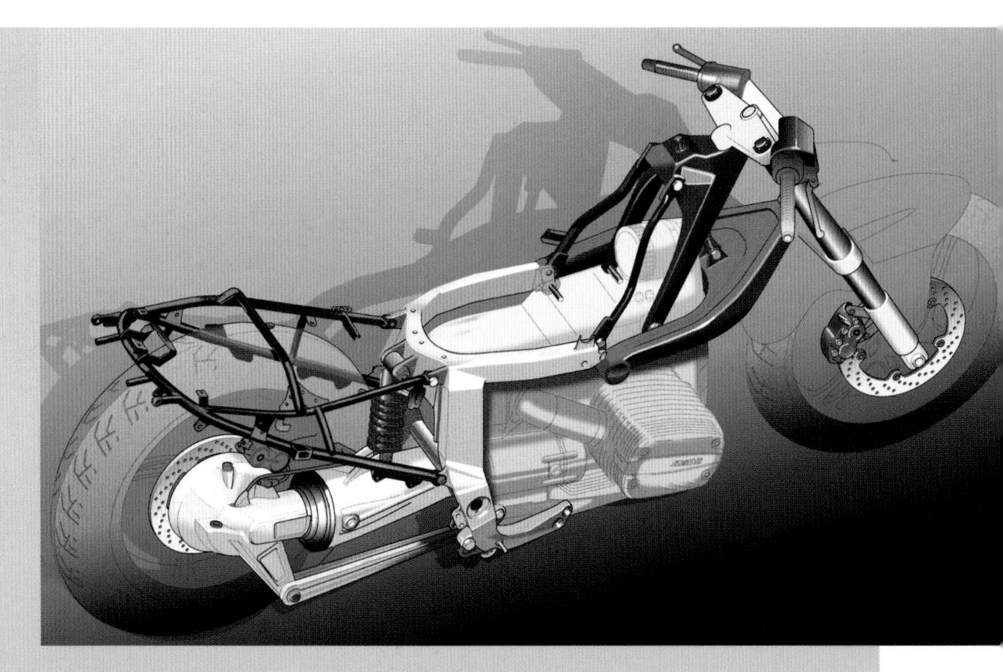

ABOVE: This cutaway drawing of the R1100 S shows the stiff aluminum main frame running from the swingarm pivot to the Telelever mounting points.

RIGHT: BMW had pulled out all the stops to make a "proper" sporting version of its Oilhead Boxer. The R1100 S was indeed much racier than any other previous Boxer, but that still left it some way off the mid-1990s state of the art in terms of Japanese sport-touring bikes.

1998–2009 K1200 LT

BMW had offered heavyweight tourers before, but the K1200 LT was the first time it had taken the leap to the super-heavyweight class pioneered by Honda's Gold Wing. In this class, all-up weights approaching half a ton are standard, and bizarre devices like reverse gears, on-board air compressors, and cup holders are de rigeur.

The K1200 LT took design cues from the company's 7 series limousines, and the fit, finish, and quality of the bodywork was exemplary. The LT design was based on the K1200 RS platform, with a lengthened swingarm and other chassis tweaks, while the 98-bhp engine had a five-speed gearbox and the electric reverse gear essential on a 378kg machine. Built-in hard luggage, a massive dashboard, luxury options including audio systems and a GPS satellite navigation—the LT had the lot and was capable enough to remain BMW's touring flagship for the next decade before being replaced by the even more stunning K1600 touring range.

TOP: The biggest **LT** to date was massive indeed, but it remained a little bit closer to a proper motorcycle than the gargantuan Honda Gold Wing. The reverse gear is operated via the small lever just above the rider's left-hand footrest.

ABOVE: It's hard to imagine now, but at the turn of the century, satellite navigation was the sole preserve of well-heeled geeks. Firms like **BMW** were at the forefront of changing this, adding **GPS** units to its cars and bikes early on. This setup from **2000** used a slightly clunky text-based interface, worlds away from today's **HD LCD** color displays, but it was pure witchcraft at the time.

BMW

The Twenty-First Century: Back in the Fast Lane

In the early part of the twenty-first century, BMW looked to be in a little bit of a rut when it came to high-performance bikes. The Big Four Japanese firms had settled into a high-tempo pattern of two-year model updates, particularly in the intensely competitive sportsbike sector. The 600cc supersport class was selling by the truckload in important markets, the Yamaha R6 tussling with the Suzuki GSX-R600, Kawasaki ZX-636R, and Honda CBR600F. Higher up the capacity classes, 750cc superbikes were being replaced by open-class liter bikes in preparation for the regulation changes in World Superbike racing. And unlimited hyperbikes like the Suzuki Hayabusa and Kawasaki ZX-12R were tugging the tiger tail of national and international lawmakers by touching 200 mph.

Bavaria still ploughed its own furrow though, and the 1150cc-engine upgrades extended across the Boxer twin range. The R1150 GS had appeared first in 1999, a 101mm bore giving the larger 1,130cc capacity and 85-bhp peak power output. That larger engine then found its way into the rest of the range. The naked R1150 R and heavyweight-touring R1150 RT appeared in 2001, with the sport-touring R1150 RS and R1150 GS Adventure launched the following year in 2002.

On the single-cylinder front, the F650 range remained a solid entry-level option, either as a gentle roadster—the 2001 F650 CS—or a harder-core off-road weapon—the 2000 F650 GS. A more extreme Dakar version of the GS appeared later in 2003, with longer-travel suspension and more dirt-focused wheels and tires, giving a modern light-middleweight option in the blooming dual-purpose/adventure sector.

The four-cylinder K range was also seeing changes, though these were more modest updates at first. The 130-bhp/266kg K1200 RS kept its place as the firm's flagship "superbike," albeit one that was nearly 50kg overweight for even the hyperbike class, never mind the sub-200kg Japanese liter superbikes. The first K1200 GT was launched in 2002. It used the laid-down engine and chassis foundations and had a larger fairing for better weather protection, more-upright riding position, better passenger comfort, and higher equipment levels. It kept the 130-bhp superbike-

A sunny day, a fast sweeping bend, and a K1200 R Sport–pure exhilaration.

R1150RS
ELECTRONIC ENGINE MANAGEMENT
BMW

level power output but was even heavier than before at 281kg. For those wanting an even more luxurious machine, the K1200 LT supertourer was still available, rounding off a K range that was advanced in technology, adequate in power output, yet woefully overweight and stuck with 1990s chassis performance.

Change was coming though. Behind the scenes, the firm's heavyweight four-cylinder K range was being totally overhauled, and the laid-down engine architecture made way for a more conventional transverse layout. The historical advantages of the fore-and-aft crankshaft in terms of gearbox packaging for shaft drive were becoming more and more outweighed by the disadvantages. It was difficult to package a modern, high-revving, short-stroke big-bore DOHC sixteen-valve four in this way; the motor ended up too long, and issues like cornering ground clearance became troublesome.

OPPOSITE: **The R1150 RS was a fairly straight upgrade from the 1100, with similar styling and chassis layout and the same mix of solid sporty performance and practicality.**

ABOVE: **The 1999 R1150 GS introduced the new 1150 Boxer motor, which replaced the 1100 engine across the range.**

LEFT: **The Dakar version of the F650 GS added 40mm more suspension travel, 870mm seat height, and a 21-inch front wheel, plus more-focused dirt tires, extended mudguard, and handguards.**

ABOVE: **A touring tweak to the basic K1200 RS, the GT added hard luggage and expanded passenger comfort in exchange for even more mass.**

OPPOSITE: **The K1200 RS was big, fast, powerful, and capable. But its engine and chassis foundations had reached the end of the development cycle, and the all-up weight in particular let it down.**

BMW knew it had to do something to address the high-performance sector. A remarkably frank 2005 press release from the UK Motorrad division was clear:

> *For many years, BMW has been well represented in most market segments, including the entry-level single-cylinder F650 range, and the flat-twin and four-cylinder touring, sports touring, cruising, adventure sport, and roadster categories. However, the high performance sports aspect of the market never saw a true representation from BMW. This presented an exciting challenge for Motorrad designers and engineers and also offered tremendous sales potential—especially in the UK with its high percentage of sports and super-sports bikes.*
>
> *Several years ago, a logical decision was made to enter the sports segment of the market and create a machine that would serve as a prestigious and exciting top-end addition to the current model line-up. The aim was for this motorcycle to attain the same level of desirability and reputation that has always been enjoyed by BMW's M-division performance cars.*

K1200RS

The new K1200 S was the result. It was a machine that, while not quite dancing to the tune set by Japanese performance engineers, worked to a very similar tempo. Listening to BMW fans and staff—who knew very well what a sportsbike was—trying to justify the compromised K1200 RS as a superbike had never been convincing. But now, BMW's UK marketing department offered a more plausible line:

"It was never BMW Motorrad's intention to create a super-sports motorcycle designed and engineered for the racetrack rather than the road—or to produce a hyper-sport machine with the highest possible top speed. The latter is no longer an important selling point among riders. BMW Motorrad's intention has been to design a machine that complements and straddles both categories—and also attracts customers from the high performance sports touring sector of the market."

These were words that high-performance motorcycle fans could believe in, and the K1200 S, with its R, GT, and R Sport variants, was a massive hit.

The K1200 range wowed journalists and brought a whole new type of rider to the Bavarian brand. But BMW had even more to offer. For 2008, it applied the magic upgrade brush across the lineup, releasing a new K1300 S, K1300 R, and K1300 GT. The engine capacity was now up to 1,293cc, with a 1mm larger bore and 5.3mm longer stroke compared with the 1200s, and power rose to a claimed 175 bhp on the S, 173 bhp on the R, and 160 bhp on the GT. The optional electronic suspension adjustment (ESA) system appeared in second-generation form, which included elastomer technology to alter spring rate as well as damping rates. An early form of traction control, ASC (anti-spin control), was available as an option, as was an official quickshifter dubbed HP Gearshift Assistant, and the BMW Integral ABS was now standard fitment.

BELOW: The first signs of change at BMW: a 167-bhp weapon of a full-bore hypersports machine, dripping in advanced technology and a total blast to ride.

OPPOSITE: The 1300 upgrade for the K1200 family was impressive: more power and torque simply dialed in and the next generation of technology added on.

BMW
K1300R

K1200 S, K1200 R, and K1200 GT

At first glance, you could be forgiven for thinking that the K1200 S was just another big BMW K-model superbike. But lurking under the large, sensibly colored fairing panels was a revolution in four-cylinder machinery from the German firm. The old laid-down motor had been dumped in favor of a more conventional layout, with the cylinders arranged across the bike, perpendicular to the direction of travel. The 1,157cc engine wasn't completely "normal"—the cylinder bank was tilted forward by 55 degrees, giving a very long, low package and putting more weight over the front wheel. It also stuck with shaft drive, unusual on such a high-performance design.

The rest of it was an untrammeled powerhouse. The top end was inspired by Formula 1 race engines, with a sixteen-valve DOHC head, short-stroke architecture of 79x59mm bore and stroke, finger-follower valve operation, forged-steel connecting rods, and a super-high 13:1 compression ratio. Lubrication was via a race-type dry-sump system, while the gaping intakes were fed by an advanced fuel-injection system with enormous 46mm throttle bodies. The result was a stunning 167 bhp from a motor that was impressively compact and light, weighing in at a claimed 81.3kg. It was even good on fuel, with a test-cycle consumption of just 51.4 miles per gallon at 75 mph.

This awesome powerplant was bolted into another novelty for the company: a twin-spar aluminum frame, not unlike the standard chassis design of most Japanese performance machinery at the time. BMW being BMW, of course, there were some unique chassis-design solutions. The Duolever

ABOVE: The K1200 engine boasted the best materials and design that BMW could muster, aimed at maintaining torque at high revs for maximum power, low friction, and a laid-down design to suit the radical chassis layout.

RIGHT: The Duolever front suspension design was based on the Hossack system: a cast-aluminum front wheel carrier mounts to the frame via a set of wishbone linkages. Springing and damping is by a centrally mounted monoshock unit.

front suspension was an all-new "Hossack" girder design, which used a cast-aluminum front wheel carrier, not unlike a vertical swingarm unit, with dual-arm linkages, a scissor-type steering mechanism, and a central monoshock suspension unit. The rear swingarm was a single-sided unit, with integrated shaft final drive and Paralever anti-torque linkages.

But it was the pure performance of the K1200 S that made it such a hit. Finally, here was a no-compromise approach to power and weight that could go toe to toe with the best from Japan, without any excuses about "smooth torque" or "mature handling." The K1200 S was a hard bastard of a bike, that 167-bhp engine propelling its 248kg dry mass almost as aggressively as the likes of Suzuki's class-topping GSX1300R Hayabusa. It had sharper handling than the Japanese hyperbikes, but it was also more practical and comfortable than the liter-class superbikes they offered. As a continent-crushing "hypersports-tourer," it excelled in a class of one.

K1200 R and R Sport

The "supernaked" class of high-powered unfaired bikes really took off in the 2010s, but BMW was well ahead of the game, launching the K1200 R in 2005. Essentially a K1200 S with the fairing removed and some radical industrial styling, it kept almost all the power, in an upright, naked roadster form. A small half-faired K1200 R Sport variant followed in 2007.

K1200 GT

The K1200 GT name had been used before, for a more touring variant of the final K1200 RS laid-down design. But BMW reused the moniker for a luxurious variant of the 2004 K1200 S. A more protective fairing, upright riding position, and even more touring practicality meant even longer-legged capability—and it kept almost all the performance of the S.

LEFT: The K1200 R was barely believable when it was launched; naked bikes just didn't have 160-bhp+ engines back in those days. Now it's par for the course, but the R was in a class of one when it launched. The R Sport was the perfect halfway point for those wanting the naked feel but with at least some chance of hanging on at top speed.

BELOW: The GT variant of the K1200 family appeared a little later than the S and R and offered a luxurious, well-appointed hypersports-touring experience for two, with full hard luggage. It was the ultimate gentleperson's express for the mid-2000s.

Electronically Adjustable Suspension

BMW raised its power and sophistication game massively with the launch of the K1200 S in 2004. And it also brought another "world first" to motorcycling: electronically adjustable suspension. Simple push-button controls on the handlebar switchgear let the rider cycle between various settings for the front Duolever and rear Paralever suspension units, altering both the damping and spring preload separately. There were three preset spring settings—solo rider, rider with luggage, and rider and passenger with luggage—and three damping settings—comfort, normal, and sport. An electronic control unit sent signals to stepper motors on the suspension units that dialed in the required settings in the same way as manual adjusters, minus the hassle and the dirty fingernails.

This initial ESA system was a big hit with the motorcycling press and riders alike. It was a much more convenient solution to altering suspension settings than having to mess about with wrenches and screwdrivers. But it was no more than that at first—convenience rather than an outright performance update. BMW expanded the usefulness of the ESA system further in 2008 with the ESA II system on the K1300 range. This iteration could actually adjust the spring rate on the rear suspension unit as if a totally different spring was mounted. A cunning elastomer cylinder above the main spring mount had a movable metal sleeve controlled by the suspension ECU. As the metal sleeve moved, more or less of the elastomer material could absorb the main spring forces, adjusting the overall spring rate as if the entire spring had been swapped for a stiffer or softer part. This is, in theory, a much more refined solution to different loads than simply adjusting the spring preload.

ESA and ESA II were popular additions to the BMW range, especially on the heavyweight touring machines and the GS models. But they were still "dumb" suspension systems in that they merely automated the manual adjustments to preload and damping that riders could make to most competitor motorcycles. The next big advance came in 2012 with the new Dynamic ESA system on the R1200 GS. This was what's known as a semi-active suspension system, where the electronic control unit is continually adjusting the suspension damper settings depending on what the bike is doing. When the rider accelerates hard, the ECU can increase the compression damping on the rear shock to reduce squat and improve traction. When hard on the brakes, the control unit can quickly increase the compression damping on the front suspension.

That simple principle—using an ECU to constantly tweak the suspension settings on the fly—can be expanded in a number of ways. You can add inputs from a sophisticated inertial measurement unit (IMU)

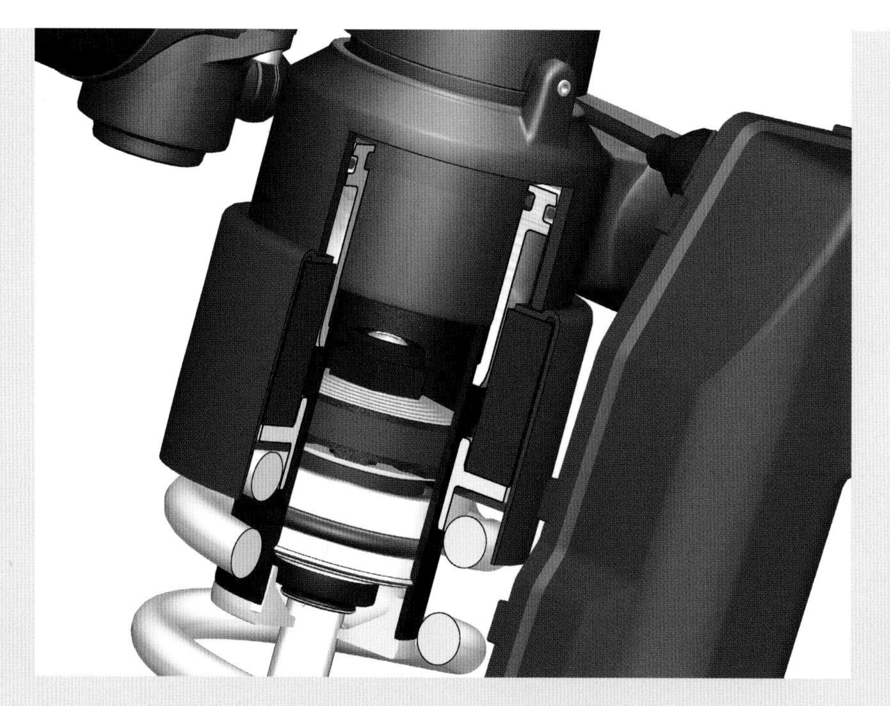

ABOVE: You can see the elastomer sleeve inside the moving metal sheath, which alters the effective spring rate as it exposes more or less of the elastomer.

LEFT: Off-road bikes are a perfect case for electronic suspension adjustment: the rider can simply dial in dirt settings from the dashboard for a fire road or green lane, then swap back to asphalt settings when back on the highway.

so the ECU knows how the bike is moving in all three dimensions in space—how far it's leaned over, if it's wheelying, and how hard it's accelerating and at what speed. You can add different algorithms for the rider to choose from—rain, sport, or track (on performance road machines)—plus enduro and dirt modes (on dual-purpose machines). You add stroke-measurement sensors so the ECU knows how far and fast the wheels are moving up and down over bumps and during weight transfer from acceleration and braking. You fit super-fast solenoid-style and stepper-motor actuators to move the damping adjusters even quicker. Do all of that and you end up with a system like the Dynamic Damping Control (DDC) setup found on the latest S1000 RR superbike: an easily accessible menu with powerful adjustability that acts like a World Superbike suspension technician under the seat, constantly tweaking the settings to cope with whatever the bike is doing thousands of times a minute.

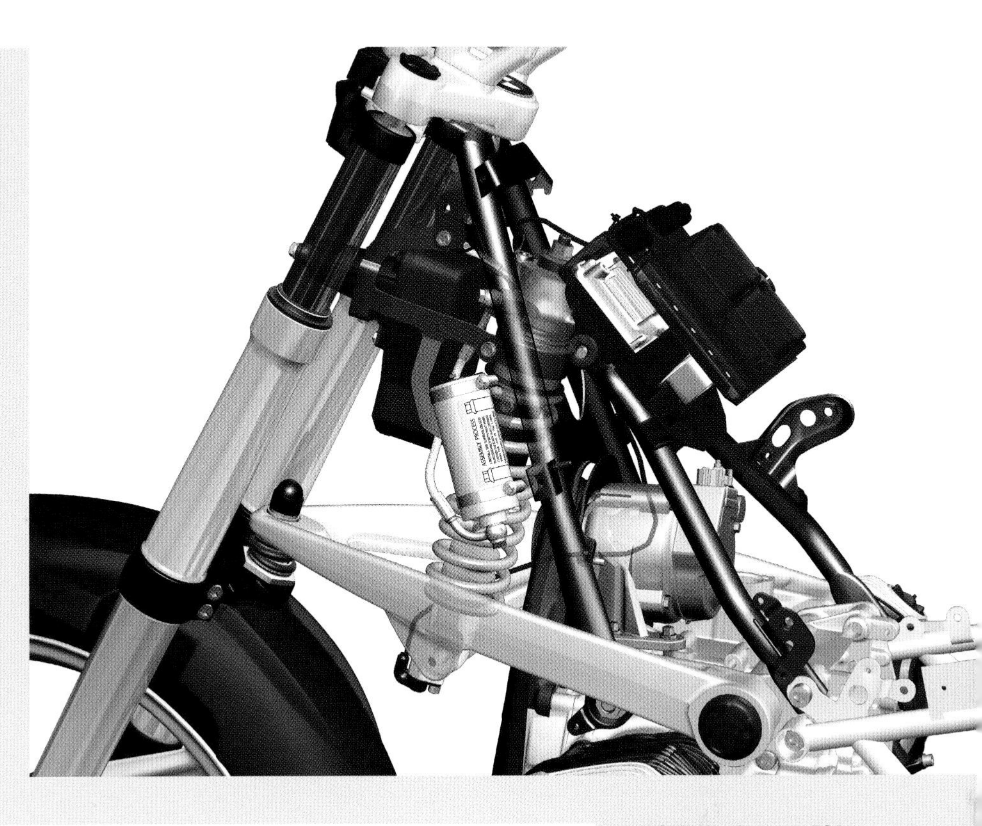

ABOVE: The Dynamic ESA setup on the 2012 R1200 GS added a semi-active function to the adjustment and used smaller, faster, lighter actuators.

LEFT: Dynamic Damping Control extended the semi-active functions out with smarter ECUs with IMUs and even higher-performance suspension adjusters.

On the road, there's little to pick between the S and ST versions of the F800. The slightly more dynamic riding position and stickier tires on the S won it for me on the riding launch in South Africa. ***James Wright/BMW***

These cutting-edge systems meant the K1300 range was among the most advanced bikes available in terms of high-tech electronic riding aids.

A New Twin for a New Century

By the middle of the 2000s, then, BMW was revitalized. Its four-cylinder machinery was on-point at the top of the market, but the middle ground needed some attention. The R850 Boxers had always been a bit of a lash-up as sleeved-down 1150s and 1100s rather than purpose-designed middleweights, and the power/weight balance was some way off. The answer was another trailblazing design choice: a totally new parallel twin engine developed with the Bombardier Rotax firm (which had helped with the F650 single) and the first new Motorrad engine layout in decades.

First released powering the 2006 F800 S and ST models, it was on the cutting edge of technology, with a compact layout, water cooling, eight-valve DOHC cylinder head, and a cunning balancer system alongside a 360-degree crank. A long "nodding" beam at the bottom of the engine pivoted from the rear of the crankcase and linked to a "slave" con rod between the cylinders. As both the pistons moved up and down together, the balancer beam traveled up and down in the opposite direction, canceling out the vibration in a satisfyingly simple, elegant fashion.

The new motor made a decent 85 bhp from its 798cc capacity and added a whole new element to the BMW Motorrad lineup. With just enough power for sporting riders and good economy and easily accessed torque for touring, the 800 twin motor soon appeared elsewhere, including the F800 R naked sport roadster in 2009. Perhaps uniquely for a BMW, the R model came about as a result of a modified stunt rider's machine. Freestyle stunt competitor Chris Pfeiffer used an F800 S in 2006, and his customized competition bike offered a template for the R model. The fairings were removed from the S, its belt final drive was replaced by a chain, the single-sided swingarm was swapped for a dual-sided unit, and a set of wider street handlebars gave a more upright riding position for better control. It made life easier for heroes like Chris Pfeiffer, but it also made for a nimble, capable urban machine for the rest of us mere mortals.

Another R-Series Renaissance

In hindsight, it's amazing just how busy BMW was in the first few years of the twenty-first century. We've seen the K-series fours totally overhauled, as well as

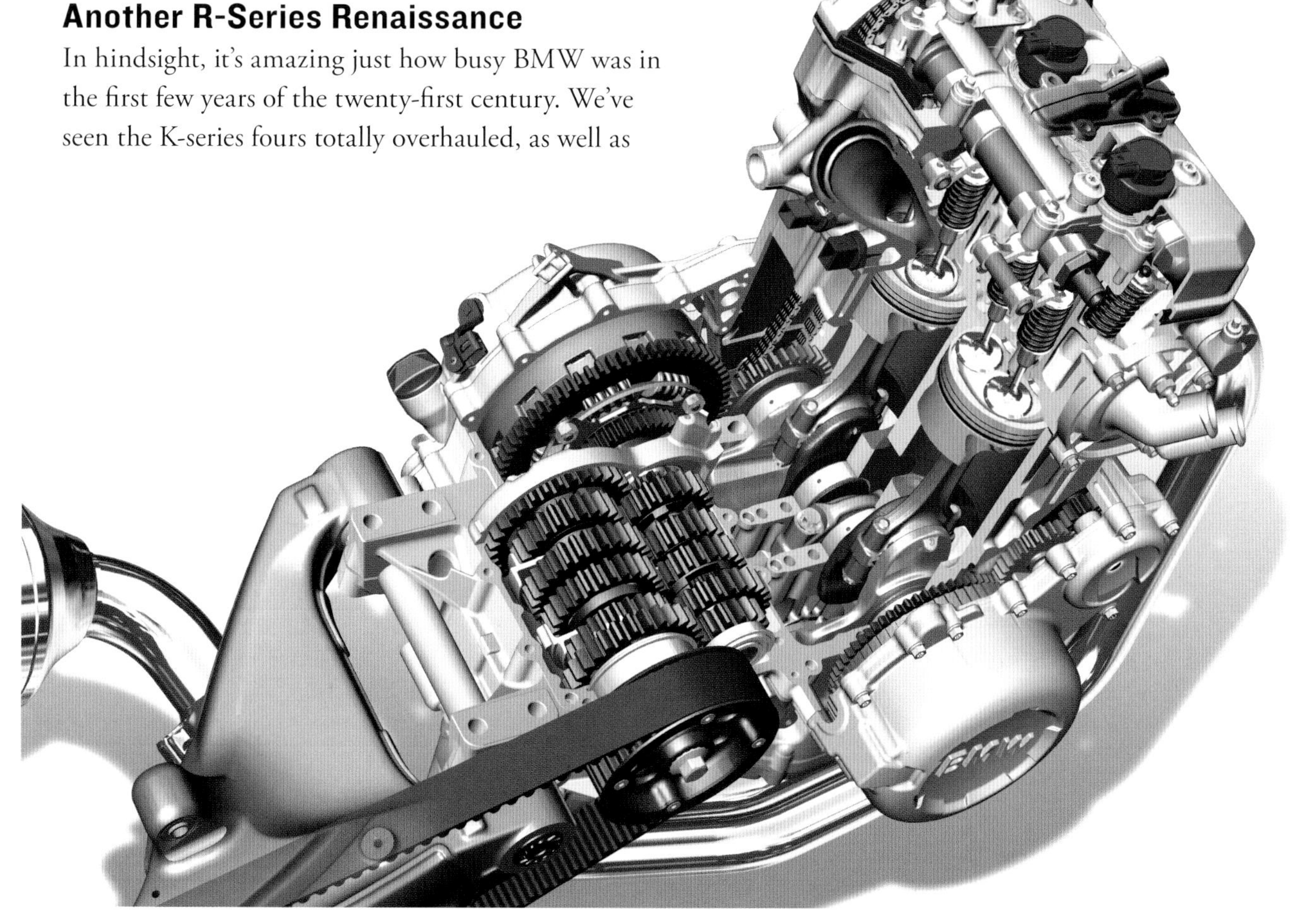

ABOVE: The central con rod is attached to a counterweighted beam that moves up and down opposite to the pistons, balancing out the vibration forces.

LEFT: The first F-series parallel-twin engine was a compact design with a lot packed in. Note the belt final drive, 360-degree crank, and "slave" balancer con rod just visible between the two main con rods.

the launch of the F-series parallel twins. But the firm's R-series Boxers were also undergoing fundamental redesigns. By 2004, the R1100 and R1150 range were polished, capable, and mature designs, which allowed riders to do almost anything from enduro riding on the R1150 GS Adventure, through pure sports riding on the R1100 S, sport-touring on the R1150 RS, and heavyweight touring with the R1150 RT.

The next step was about to come, featuring an all-new 1200 engine plus extensive revamps of chassis setups, a big weight-loss program, and further ratcheting up of technology and equipment.

There was a hint of where the direction of travel might be heading already—BMW had actually been selling a bigger 1200 Boxer since 1997. The R1200 C custom cruiser had been a modest success in several markets, and it was that bike's larger 1,170cc engine capacity that found its way into the first of the new generation of 1200s, the R1200 GS, launched in 2004. The new engine used the same longer-stroke layout

BELOW: **The 1997 chromed custom cruiser C variant was the first Boxer to sport the 1,170cc capacity used across the range from 2004 on.**

OPPOSITE: **If you'd suggested that BMW would offer factory support to a stunt rider in the 1990s, you'd have gotten some strange looks. Chris Pfeiffer helped transform the firm's image through the mid-2000s.**

Red Bull
Red Bull
KNAUS
AKRAPOVIĆ
CHRISPFEIFFER.COM

F800 S and F800 ST

All the capability of BMW's bigger machinery in a smaller, lighter package? The new F800 S and ST had plenty to live up to when they launched in 2006. On paper, they were a little pedestrian in terms of power and mass: 85-bhp peak power and 204kg wet put them above the likes of Suzuki's SV650 V-twin roadster but below the Yamaha Fazer 600 or Honda Hornet. The belt final drive lost a few cool marks, but the single-sided rear swingarm snatched them right back. Brembo four-piston brakes up front were from the upper end of the market, a proper aluminum-beam frame gave stiffness without excess mass, and the neat dual-clock dashboard had a very useful LCD display panel alongside the analog speedometer and tachometer.

The European press riding launch took place in South Africa, near Cape Town, and I was invited to cover it for *SuperBike* magazine. A long day on the roads from Franschhoek to the coast showed a capable, nimble machine with decent sporting abilities and loads of practical features. The ST had a slightly larger fairing with lower side panels, different wheels, and a luggage rack, and included panniers in many markets. Both the S and ST had an unusual underseat fuel tank, with a rather parsimonious 16-liter capacity that reduced ultimate touring range a little. But the overall package more than lived up to the initial promise. The optional ABS, extensive trip computer functions, heated grips, and factory-option luggage did indeed make for a practical, comfortable, high-tech machine, but with much less weight than BMW's bigger touring bikes. It was also great fun: the handling was precise and agile, the suspension provided plush wheel control and ample feedback, the engine gave great drive out of corners, and the brakes were more than up to the job.

Perhaps the most interesting use of the 800 twin motor came in 2008, when BMW released the adventure-styled F800 GS. This filled the gap between the lightweight single-cylinder F650 GS and the heavyweight R1200 GS, formerly catered for by the relatively rare R850 GS. But the F800 was a much better bike than the underpowered, overweight 850 Boxer, with weight and power levels more suited to a wide range of riders. It was arguably the first "modern" twenty-first-century middleweight adventure bike, treading a path taken later on by firms like KTM, Triumph, Yamaha, and Ducati, and was a big hit straight away.

Confusingly, BMW also replaced the single-cylinder F650 GS with a detuned, lower-spec version of the 800, called, er, the F650 GS. The 2008 F650 GS twin was much more road-biased than the 800 and used a variant of the 798cc engine that made 15 bhp less peak power. It came with conventional rather than USD front forks, less suspension travel, lower seat height, and a smaller 19-inch front wheel, rather than the 21-inch rim on the 800. The wheels were cast aluminum rather than wire-spoked, tires were more suitable for road use, and the front brake had just one disc rather than the twin discs of the 800.

The single-cylinder F650 GS was discontinued, only to appear later on in the G650 range. The lesson here is that if you're buying an old F650 GS, check the number of cylinders before you hand over the cash.

ABOVE: A new breed of middleweight sport-tourer for BMW, the F800 S was the first bike to use the innovative parallel-twin engine design.

LEFT: The ST was largely identical to the S, with differences restricted to a larger fairing, different tires, a luggage rack and pannier fittings, plus wider handlebars. *James Wright/BMW*

OPPOSITE: It was perhaps inevitable that the F800 range would spawn a GS version before too long. There was a fairly extensive overhaul of the chassis from the S and ST designs and a solid off-road bias to the performance.

F800
GS

BMW

F800 GS

The F800 GS engine was a variant of the same 85-bhp, 798cc twin as the previous F800 S and ST models, with the cylinders repositioned to a more upright angle. The chassis was extensively overhauled for dual-purpose use, including longer-travel dirt-friendly suspension, more upright riding position, off-road wheel and tire sizes (including a "proper" 21-inch front rim), and chain final drive rather than belt. BMW also swapped out the aluminum twin-beam frame for a tougher steel-tube fabricated design, while the single-sided swingarm off the S and ST was replaced with a conventional dual-sided arm. Dry weight was an impressive 178kg, and with the 16-liter underseat fuel tank filled and ready to ride, the 800 GS weighed 207kg.

Optional crash bars and hard luggage provided even more off-road styling and touring ability, and even fully kitted-up for continent-crossing adventure, it remained slim and manageable.

A steel-tube trellis frame replaced the alloy beam frame on the F800 S/ST for better resistance to the rigors of off-road riding. A new dual-sided alloy swingarm and chain final drive, plus long-travel upside-down forks, completed the GS chassis revamp.

as the 1200 C, but the rest of the motor was totally revised, making a claimed 18 percent more power, now up to 100 bhp from 85 bhp. The transmission gained a sixth gear for more relaxed high-speed work, while the Paralever shaft drive was a new design, with the torque arm above the shaft for extra ground clearance and less chance of damage when riding off-road.

The running gear was similar to the 1150 in design but totally different in execution. It still used Telelever front suspension and a steel-tube trellis-type frame, with the engine as a structural foundation. But the whole chassis had been thoroughly revised with weight-saving in mind, and together with the lighter engine, BMW shaved an incredible 30kg in mass off the big enduro Boxer.

That big boost to the power/weight ratio made a tremendous difference to the performance of the GS in a straight line. But it arguably had an even bigger effect on the feel and agility, both on- and off-road. The 229kg R1150 GS did a great job of disguising its hefty nature most of the time. But with the new

OPPOSITE: The lighter, stronger, smarter R1200 took the flagship GS range to a new level for 2004.

ABOVE: The best part of the 1200 GS was the weight loss: careful design around the motor and chassis lost a total of 30kg overall.

2004 R1200 GS

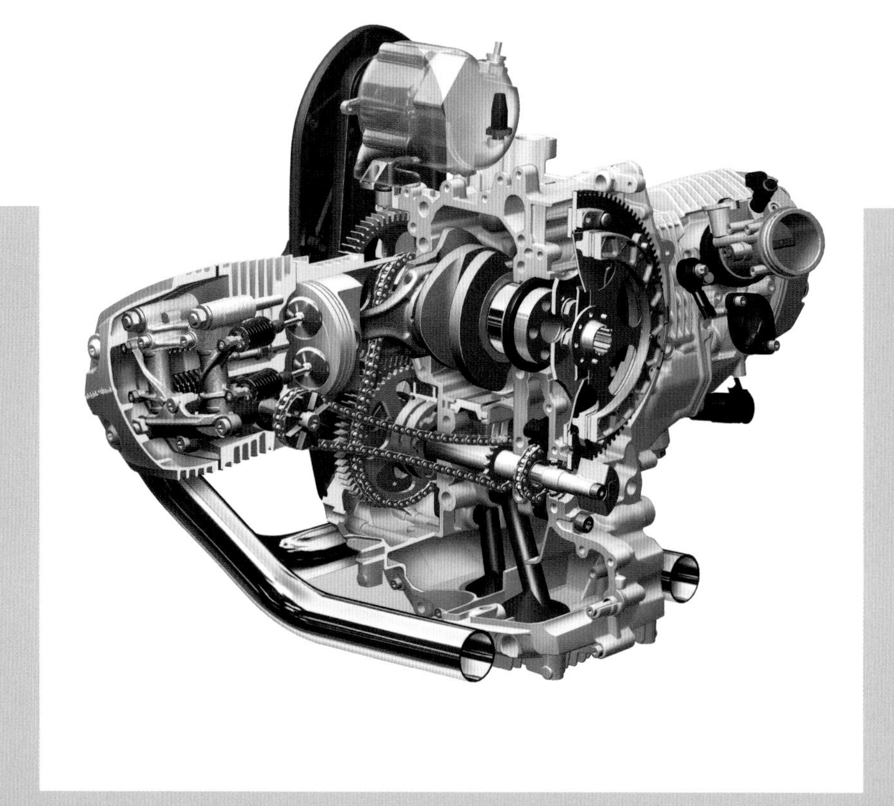

The R1200 GS was a total redesign of the R1150 GS, with every part revised and upgraded. High-end chassis design technology helped save 30kg overall, a new EVO brake system with integrated ABS gave even better stopping performance, and an advanced CANBUS electrical system improved reliability and reduced wiring. The large 20-liter fuel tank was made of lightweight plastic, a new top fairing featured an adjustable windshield, and BMW offered a huge range of optional accessories, from heated grips to crash bars to hard and soft luggage.

But it was the new engine that was the most significant upgrade, saving 3kg and producing 100-bhp peak power—15 bhp more than the 1150. The bottom end now featured a balancer shaft for the first time to reduce the inevitable second-order vibrations from the Boxer layout. The new crankshaft was more compact and 1kg lighter, despite the 1200's longer stroke, while the dry clutch used a larger diameter friction plate. Three-ring pistons were 10g lighter than before, despite the 1200 using the same 101mm bore size, and the cylinder heads were fitted with 2mm larger inlet and exhaust valves. The "high cam" design remained, with short tappets and rocker arms rather than true overhead camshafts.

The engine management system took a significant leap forward, both to boost power and to cope with the ever-tightening demands of European emissions regulations. The BMS-K system operated a fully sequential fuel-injection program, with new knock sensors that helped the engine run efficiently with its new, higher 11:1 compression ratio, even on lower-octane gasoline. Finally, an advanced three-way catalytic converter in the lighter stainless-steel exhaust system virtually eliminated CO and HC emissions.

It was huge progress for the venerable BMW Boxer engine—more power and torque, less weight, and a new six-speed transmission mated to an overhauled, lighter shaft final drive. It helped the GS further cement its reputation as the best of the large adventure touring machines and kept it at the top of the sales charts across Europe.

ABOVE: The 1200's new balance shaft is cunningly mounted below the crankshaft, concentric inside the existing auxiliary shaft, which drives the camshafts and oil pump.

RIGHT: The first R1200 GS prototype used the 1150 as a base.

1200, you had 100 bhp of grunty flat-twin power and a dry weight of just under 200kg, meaning much less weight to try and cover up during road riding (and less to pick up when it toppled over on the dirt).

The new 1200 Boxer was rolled out to the rest of the Boxer range over the following years, starting with the full touring R1200 RT in 2005, then the sportier R1200 ST. The following year saw the 2006 R1200 GS Adventure extend the adventure range upward while the R1200 S replaced the R1100 S sportsbike. Finally, in 2007, the R1150 R was replaced by the updated R1200 R. Alongside these updates came a new family of super-trick Boxers—the HP2 range.

ABOVE: **The ST had a face only a mother could love, but under the curious fairing design lay a solid sport-touring Boxer with the new 1200 motor, fine handling, and extensive equipment levels.**

RIGHT: **The 1200 maintained the RT's role as BMW's heavyweight techno-tourer for mileage junkies. It was also a common base for police and other emergency service users.**

BMW's dalliance with full-fat competition machinery in the form of the **G450X** enduro machine was sadly short-lived. The bike gained many fans in its short life, and advanced tech like the chain drive arranged concentrically with the swingarm pivot and crank-mounted clutch was far ahead of its time.

The R1200 GS and F800 GS weren't the only new dirt-focused machines to wheelie their way out of the Spandau factory doors in the 2000s. BMW came up with a huge range of new off-road bikes, from the sublime G450 X enduro racebike to the ridiculous HP2 Enduro. The F650 single also reappeared, rebranded as the G650 and produced in China.

The G450 X was perhaps the biggest surprise in terms of new metal; a very serious-looking track-ready enduro, it produced 50 bhp from its unique engine design. Radical out-of-the-box tech was everywhere, from the clutch on the end of the crankshaft to optimize the frame tube position to the swingarm pivot that was concentric with the final drive sprocket.

Away from the showroom floor, BMW also made a big corporate move in the dirt world. In 2007, it bought the Swedish Husqvarna Motorcycles brand from the MV Agusta/Cagiva Group. Speculation immediately ran wild, with industry figures predicting a whole new range of BMW/Husqvarna mashup machinery. Sadly, the financial crash of 2008 hit hard, and the Husqvarna partnership bore no fruit at all. BMW sold the brand to KTM in 2013, where it was used as a boutique sub-brand for urban road machines and high-end dirt bikes.

Munich on Top at the End of a Busy Decade

By 2009, life was looking pretty good for BMW. Its model lineup was a world away from the slightly tired, overweight, underpowered range of heavyweight machinery of ten years before. The new K1200 and 1300 range could stand alongside the best of the Japanese competition, while the R1200 GS and GS Adventure were amassing enormous sales volumes and arguably building a whole new sector—the big adventure touring bike—all on their own. The midrange F800 machines weren't setting the world on fire in quite the same way, but the F800 GS in particular was again defining a new class of bikes, the upper-middleweight adventure tourer/enduro machine.

The rest of the Boxer range remained, perhaps, more typically BMW; curiously styled oddities like the R1200 ST and the R1200 CL still stood out against the mainstream. But they also just worked really well. If you had to cross Germany and France or travel the length of the US West Coast, a BMW R1200 RT would make an extremely efficient, luxurious job of it.

But there were some other factors working in the favor of BMW and other European manufacturers. The financial crash of 2008 had had a particularly heavy impact on the Japanese motorcycle industry. It faced the same credit crunch and lack of demand as everyone else while also dealing with the massive

ABOVE: **For riders who didn't need the sheer size and weight of the R1200 GS, the F800 GS offered much of the same capability in a smaller, more easily managed package.**

LEFT: **Someone at BMW clearly had a bet on here: the R1200 CL fairing is spectacularly lacking in conventional beauty. Once you sat behind it, though, you didn't mind. Luxury touring kit including a stereo sound system and cruise control made this version of the R1200 C one for the long run.**

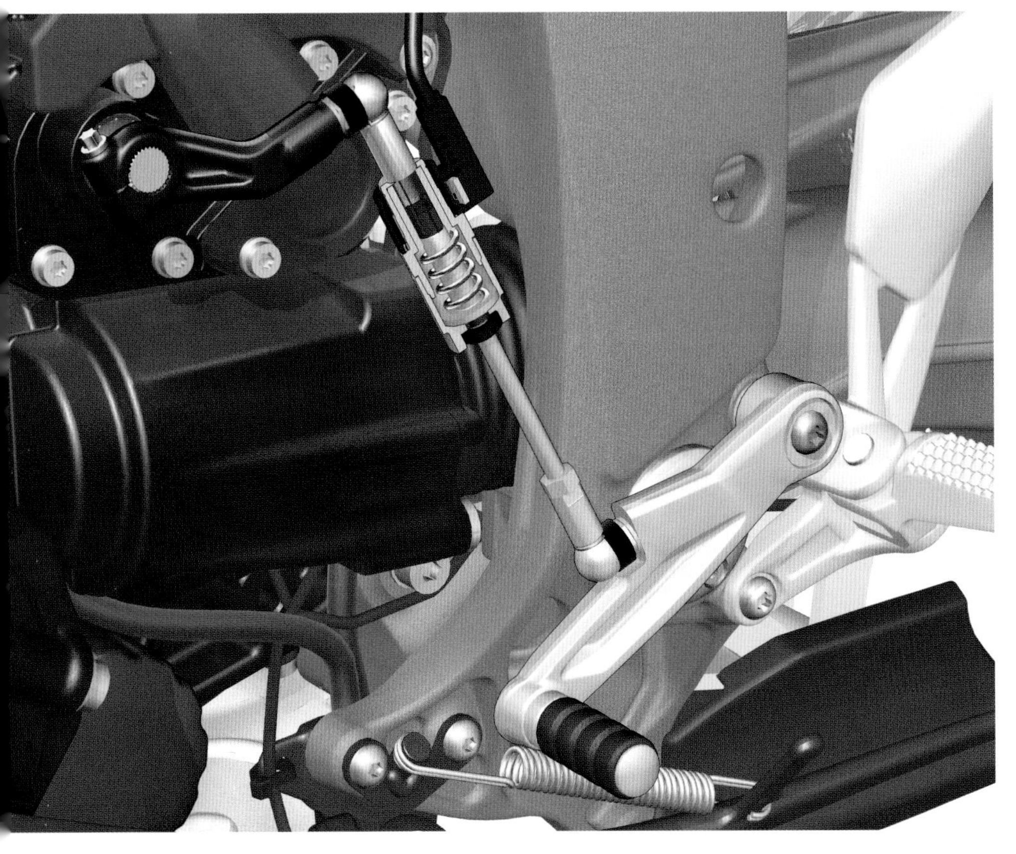

increase in the value of the yen. Suddenly, new Japanese machinery was much more expensive to buy in Europe, Australia, and the United States. For example, in late 2009, a Yamaha FJR1300 ABS cost £14,499 in the UK, but the BMW K1300 GT was just £12,525. The Japanese firms worked hard to keep their prices competitive, but in general, the historical price premium that owners had expected to pay for a BMW was eroded or disappeared altogether.

Meanwhile, the traditional BMW brand values stayed in place—great dealer and aftersales backup, higher resale prices, and good equipment levels. It wasn't all smooth sailing, as the firm had a number of production and quality issues stemming from new technology in the late 2000s. One particular problem lay in a new type of switchgear design, where the electrical circuits were "printed" onto the body of the switch housings themselves instead of separate circuit boards. This was lighter, more elegant, and potentially cheaper, but early production suffered from unreliable performance, and several models had to be recalled.

There were also problems with ignition switches on some models. As ever, how a company deals with a problem is more important to consumers than the problem itself, and BMW earned plaudits for the speed and ease of its recall operations.

These big changes in the market weren't restricted to the Munich firm. KTM, Ducati, and Triumph were all snapping at the heels of Japan's Big Four. Kawasaki, Honda, Yamaha, and Suzuki had arguably rested on their laurels a little and now lagged behind on technology in the advanced, super-competitive markets of Europe and the US. BMW's work with ABS, traction control, electronic suspension, gearbox quickshifters, LCD dashboards, and trip computers were miles ahead of the offerings from the Far East.

But there was more to come—much more. Back in 2007, rumors had been spreading through the bike industry about a team of BMW engineers who were spending *a lot* of time and money hiring out racetracks for private test sessions. And what had been seen sneaking out of trucks and into closed pit garages? Only the very best Japanese 1,000cc superbikes, including the legendary Suzuki GSX-R1000 K5/6 model. There were even whispered stories about unusual-looking machines, with GSX-R1000 bodywork covering decidedly non-GSX-R chassis and engine components. Just what were they up to?

The answer came in 2009 with the release of the S1000 RR superbike, a no-holds-barred assault on the outright sportsbike performance crown—as well as a base for the first WSBK and BSB racebike from BMW. It matched—and exceeded—the spec panels of the very best liter superbikes that Japan could offer: the Honda CBR1000RR Fireblade, the Yamaha YZF-R1, the Kawasaki ZX-10R, and the Suzuki GSX-R1000 the Bavarian engineers had been benchmarking a few years earlier. Europe was coming, and BMW was in the vanguard.

OPPOSITE TOP: This carbon-fiber-clad S1000 RR prototype racebike is dripping in the very best of kit: OZ wheels, Öhlins gas forks, Brembo WSBK brakes, Pirelli slicks, and Akrapovic race exhaust. It's a very serious statement of intent.

OPPOSITE: The K1300 brought another first for BMW: a factory-fit quickshifter. A switch on the gear-change lever killed the motor for a split second as the rider's foot moved the pedal up. That cut torque to the gearbox, allowing the selector forks to change to the next-higher ratio while keeping power interruptions to a minimum.

LEFT: Finally, with the S1000 RR, BMW had a genuine top-end superbike, with stunning performance that could take on the best Japan or Italy could offer, on road and track.

The HP2 Family

At times in the early 2000s, it seemed like BMW was in danger of letting its long-term relationship with the Boxer-engine layout turn into a damaging obsession. The R1100 S was a proper sportsbike according to the firm. But it was massively overweight (208kg dry) and underpowered (98 bhp) for the sportsbike class at the turn of the century, and the Telelever/Paralever shaft drive chassis setup couldn't match a supersports race replica's fork, monoshock, and chain drive chassis.

Once again, BMW's stubbornness in sticking to its engineering principles was a sight to behold. First, the firm brought out an R1200 S in 2006, with the bigger engine from the recent GS. Launched at the Killarney race track in Cape Town, South Africa, the 1200 S was better than the 1100, of course—the Boxer engineers had cut nearly 30kg from the weight somehow and tuned the 1,170cc flat-twin oil-cooled motor up to a claimed 122 bhp, lifting the R1200 S firmly into the realm of Ducati's 1000 SS, another old-school big-bore twin sportsbike.

The real madness was yet to come. First, BMW cranked things up one last time on the R1200 S to produce the HP2 Sport. HP stood for high performance, and the 2 stood for the twin-cylinder engines. There were two other HP2 models as well: the HP2 Megamoto and the HP2 Enduro. All made more power than stock, up to 128 bhp on the Sport, which had a special DOHC cylinder head design, and all had race-spec suspension and chassis parts. They also had stunning design; the Megamoto in particular looked completely insane, with super-long suspension travel, carbon bodywork, Öhlins rear shock, and a hand-built Boxer engine.

OPPOSITE TOP: The HP2 Enduro added a pure off-road vibe to the HP2 family, thanks to proper race-spec Öhlins off-road suspension, full dirt tires and wheels, and a single front brake disc.

OPPOSITE: The 128-bhp HP2 Sport was much more exciting than the 1200 S. Carbon bodywork, the most powerful Boxer motor ever at the time, and premium chassis kit made a beguiling mix.

ABOVE: The R1200 S was a solid upgrade on the slightly disappointing R1100 S, with more power, less weight, and decent chassis tech. *James Wright/BMW*

SHOEI
HAWK
RACING
ROCK OIL
6
6
BMW Motorrad
BUILDBASE
BUILDBASE

Racing

In this history of the BMW Motorrad century, we've focused mostly on the firm's road bikes. But the company also has a long, if slightly unconventional, history in competition. It started out in the very early days when bikes like the R37 and R47 were used in pioneering motorsports events by Rudolf Schleicher and Ernst Henne. Then, in the 1930s, BMW made a series of land speed record attempts with special supercharged Kompressor streamliner bikes. Henne set a new world speed record of 173.7 mph (279.5 km/h) on the autobahn near Frankfurt am Main on November 28, 1937, a record he would hold for the next fourteen years. The firm also enjoyed success at the 1939 Isle of Man TT, where Georg Meier became the first non-Brit to win the Senior race on a Type 255 Kompressor machine. These efforts were urged on by Nazi propagandists to show the power and success of the new German Reich.

It's probably fair to say that the Munich firm underachieved in "normal" bike racing in the second half of the twentieth century. A dispassionate observer would, of course, make allowances for the chaos and devastation of the Second World War. But compared with Japanese firms, who barely had any prewar bike industry to speak of and whose manufacturing base was even more comprehensively destroyed, BMW was largely missing at the highest levels of solo bike racing. Yamaha, Honda, Suzuki, and Kawasaki devoted huge chunks of their initially meager resources to compete against the British and Italian giants in blue riband Grand Prix racing. Meanwhile, BMW was pottering at the edges with incredible sidecar racing success at the highest level but little in the way of solo performance at world level. BMW racer Walter Zeller is the main standout, finishing second in the 1956 500 World Championship on an exotic RS54 factory racebike.

A Series of Sporting Handicaps

It seems clear now that it was BMW's exclusive shaft-drive Boxer engine powertrain policy that hampered the firm in much competition. Early in prewar road racing, the more conventional single- and twin-cylinder British machines had better handling, but the blown BMW had stronger power. And when BMW's supercharged engines were banned in most race classes after World War II, the power advantage of the Kompressor bikes

Michael Dunlop wins at the TT in 2014.

ABOVE: **Henne at Karlsruhe**

OPPOSITE: **Georg "Schorsch" Meier was already a highly successful bike racer, winning the 500cc European championship in 1938. But he reached another level in 1939, when he became the first non-British rider to win the Senior TT.**

evaporated. Later in the 1960s, when multicylinder inline racing engines appeared from Honda and MV Agusta, they had much more power too. By the time the Japanese moved on to two-stroke engines, the pure horsepower performance gap had become an abyss.

BMW's shaft drive was a constant handicap too: back then there were no clever Monolever or Paralever back ends, and the chain-driven competition enjoyed another advantage in terms of weight and handling. The move to Earles front ends in the 1960s further eroded the track abilities of BMW's bikes. As a result of all this, BMW has played a very small part in solo Grand Prix world championship motorcycle racing since the modern GP series launched in 1949. That might seem like a bit of a surprise to modern readers used to high-performance BMW bikes like the S1000 RR superbike.

Seventies Superbike Success

By the 1970s, though, BMW's bikes were improving rapidly in terms of performance. First the R75/5, then the R90 S, combined strong engines with well-designed chassis setups. The R90 S in particular was close enough to the state of the art that in the right circumstances, it might be competitive. With a highly motivated race team and top rider—plus a little bit of luck—BMW's flagship sportster could be a contender.

Those circumstances all came together in the United States in the mid-1970s. The US importer of BMWs, Butler and Smith, had been taken over in 1971, and the new owner, Dr. Peter Adams, was keen to give a more dynamic image to the brand. The US arm started with the R75 /5, building it into a 750 superbike racer with factory chassis and engine parts before moving onto the R90 S when it was released in 1973. The top rider arrived when Butler and Smith signed up British racer Reg Pridmore, who'd moved to California in the 1960s.

The Type 255 looks every bit the hard-core road racer here. Note the supercharger pipework below the cylinder, feeding pressurized air back to the intake ports. Plunger suspension is better than an unsprung rear, but it's less sophisticated than proper swinging-arm setups.

The piece of luck was the AMA superbike race series, which launched in 1976 with a heavyweight production class aimed at modified 1,000cc street bikes. The Japanese four-cylinder superbikes looked like a shoe-in for the top of this class, but for the first season, European twin-cylinder bikes from Ducati, Moto Guzzi, and BMW were on top, winning every race. Lighter weight, better handling, stronger braking, and the long race experience of the Butler and Smith team, together with a wily rider in Reg Pridmore, beat the sheer horsepower of the Japanese bikes, and BMW took the inaugural 1976 AMA Superbike title on the R90 S. The Japanese learned their lesson in 1976 and worked hard on the chassis and engine designs on their inline-four superbikes. Japanese firms would win the next fifteen titles before being beaten by another European twin—Ducati's 916—in 1993 and 1994 (though Japanese bikes have won every AMA superbike title since).

Island Life

The 1970s also saw a return to solo Isle of Man TT success for BMW (it had utterly dominated the sidecar TT class through the 1950s and 1960s, of course). Nearly forty years after the 1939 Senior TT win, BMW tasted success on the Isle of Man again, this time in the 1976 Production TT. Hans-Otto Butenuth and Helmut Dähne rode the BMW R90 S to 1,000cc class victory in the Production TT, finishing in fifth place overall on the road behind Yamaha, Suzuki, and Honda racebikes in the smaller

Record-Breaker

To go fast, you need two things: plenty of power and minimal drag. The force from air resistance increases geometrically, meaning that to double your speed, you need eight times the power, all other things being equal. Reducing that drag as much as possible is even more important than building a very powerful engine.

In the 1930s, BMW could build a powerful bike engine. Supercharging let its engineers produce 60 bhp from the Type 255 Kompressor 500cc Boxer twin. But to top the land speed record of around 150 mph, it also needed a very aerodynamic bike. Ernst Henne's 1937 bike used a special streamliner fairing, which looks more like an airplane fuselage than a motorbike. Long, smooth side panels, a bullet-shaped nose, and tail-fin back end helped Henne cut through the air and top 173 mph with just 60 bhp from that supercharged engine.

TOP AND ABOVE: Ernst Jakob Henne's streamliner bike shows the long, slippery bodywork design necessary for setting speed records. Note the air intakes at the front needed to keep the supercharged Boxer engine cool, and the outrigger wheels needed when stopped. The tail fin at the rear provides straight-line stability at speed, and even Henne's helmet is specially shaped to reduce drag.

Stunning Sidecar Success

The chassis disadvantage of BMW's exclusive shaft drive policy was much less important in sidecars, as was the wide engine. If your vehicle doesn't lean over, you don't have to worry about grounding out a Boxer cylinder head. The adoption of the Earles fork as standard was a big advantage for sidecar use. As a result, the firm dominated the three-wheeled racing world in the 1950s and 1960s. The Boxer engine from the RS54 solo racebike made decent power and was reliable, even while using an early mechanical fuel-injection system. Overseen by BMW's motorsports supremo Alexander von Falkenhausen, BMW's old-school four-stroke-powered sidecars won nineteen out of twenty titles between 1954 and 1974, before the two-stroke-engine era ended their dominance. Like Honda (initially), BMW refused to go down the two-stroke race engine route in the 1970s and couldn't compete against the super-light, super-powerful 500cc two-stroke engines from Japan.

TOP: The rather primitive nature of 1950s racing sidecars is shown in this 1953 action picture of Wilhelm Noll and Fritz Crohn.

ABOVE: By the 1970s, the modern streamlined, integrated sidecars were beginning to emerge. This 1972 picture shows Klaus Enders and Ralf Engelhardt on a sidecar chassis built by Dieter Busch.

classes. Rather like the AMA series win, it would be a one-off victory, with class changes at the TT and rapidly improving Japanese superbikes quickly overtaking the R90 S.

TT Single Success

When BMW launched the F650 Funduro in 1993, no one could have predicted that the next Isle of Man TT success would come from this soft dual-sport machine. But that's exactly what happened. British rider Dave Morris took the F650 engine, tuned it, and bolted it into his own aluminum-framed chassis design, manufactured by Harris Performance. Morris won the TT Singles race three times on his BMW-backed Chrysalis Racing Team machine in 1997, 1998, and 1999, before his tragic death at Croft race circuit in September 1999.

Another Senior TT Win, Seventy-Five Years after the Last

The Type 255 Kompressor was more than capable of winning races before World War II. But setting the valiant efforts of the R90 S aside, it was seven decades before there was another BMW

TOP: The Butler and Smith race team had been producing fast Boxer racebikes for five years before winning the AMA title in 1976. Here Reg Pridmore rides one of the special F750-class R75/5 machines built in 1971.

ABOVE: Reg Pridmore (163) leads fellow BMW racer Steve McLaughlin (83) on their BMW R90 S racebikes in the 1976 season-opening Daytona 200. Pridmore finishes the season top of the pile, becoming the first AMA Superbike champion on the Butler and Smith R90 S.

20
20

bike with the performance to dice with the very best competition on track. That bike was the S1000 RR, first seen in 2009, seventy years after Georg Meier's Senior TT win. And just five years later, the S1000 RR would match its ancestor at the Isle of Man, winning the 2014 Senior TT under road race legend Michael Dunlop. Dunlop (nephew of Joey Dunlop, brother of William Dunlop, son of Robert Dunlop, all racers) had dominated the 2013 TT with four wins in race week. In 2014, he repeated that incredible achievement, but this time he took three of those wins—the Superstock, Senior, and Superbike titles—on the Hawk Racing BMW S1000 RR. It was a stunning success for BMW and confirmation that the firm's new superbike really could win at the top.

It was a very welcome victory. The RR had done well enough in short-circuit superbike racing but hadn't had the breakthrough success some had expected. Marco Melandri had shone in the 2012 WSBK season, finishing the year in third behind

ABOVE AND LEFT: **Michael Dunlop swept the board in the big bike classes with his Hawk Racing S1000 RR at the 2014 TT, winning the Senior, Superstock, and Superbike races.**

OPPOSITE: **The R90 S had an incredible year in 1976; it took a class win in the Isle of Man Production TT ridden by Hans-Otto Butenuth and Helmut Dähne. Dähne is shown here muscling the beefy Beemer around the Mountain Course.**

SHOEI
Bathams
BREMONT
MGM
DUNLOP

Max Biaggi on the Aprilia RSV4 and Tom Sykes on the Kawasaki ZX-10R, with six wins. Melandri dropped back a spot in 2013, finishing the season in fourth place, before switching to Aprilia for 2014. BMW's WSBK efforts fell back further over the following years, with a few more top-ten spots during the rest of the 2010s but nowhere near the dominance of Kawasakis, Ducatis, and Yamahas.

Despite its short-circuit shortcomings, the S1000 RR has maintained a superb record in road racing. It holds the current TT outright lap record of 135.452 mph under Peter Hickman, who also holds the fastest newcomer lap and Superstock lap records at the Isle of Man, and won the Superbike and Senior double at the 2019 TT. Hickman has also campaigned the S1000 RR in British Superbike and at the Macau GP street race, which he's won three times on the BMW—in 2015, 2016, and 2018.

BMW redoubled its efforts for superbike racing in 2021, launching an even more hard-core homologation version of the S1000 RR, the M1000 RR. With carbon-fiber winglets on the front-fairing carbon-fiber wheels plus titanium Pankl rods, and exotic two-ring race pistons inside the motor, the M1000 RR has been designed as an uncompromising assault on superbike and superstock racing, taking on the likes of Ducati's Panigale V4 R. Since the high-end race parts are fitted to the showroom bike, they're automatically eligible for production-based race series, giving extra latitude to tuners and race team managers.

ABOVE: **The latest S1000 RR is a monster of a superbike as it is, but the M 1000 RR moves it to another level. Note the less exotic suspension parts: these are the first things replaced by race teams, so there's no point fitting high-end road kit to this race homologation special.**

OPPOSITE: **The unforgiving nature of the armco-clad Macau street circuit is seen here, as Peter Hickman wrestles his S1000 RR around in 2019.**

FOLLOWING PAGES: **In 2012, Marco Melandri had BMW's best WSBK season so far on the S1000 RR. Here he's following his teammate Leon Haslam around Brno at the Czech round, where Melandri won both races that year.**

Arai
BECKER
CARBON
alpha
RACING

33

Paris–Dakar Rally

The Paris–Dakar Rally is another unconventional episode in BMW's diverse racing story. The firm was a strong contender in the very earliest of off-road racing, competing at the top of the International Six Day Trial (ISDT) events in the interwar period. But in the 1950s and 1960s, as in the road race world, competitor machines without the handicap of shaft-drive Boxer powerplants took over. A skinny, light, single-cylinder scrambler with chain drive and conventional long-travel suspension had a definite edge over an R-series BMW twin for most off-road riding. The /5 machinery of the 1970s offered fresh possibilities though, and BMW enthusiasts around the world modified their R75/5s for the dirt, fitting high-level exhausts, long-travel forks, and rear shocks, and swapping in off-road wheels and tires.

The late 1970s saw an exciting new off-road race—the Paris–Dakar Rally. Launched by desert racer Thierry Sabine in 1979, it did what it said on the tin: it was a race for cars, bikes, and trucks, from Paris, France, to Dakar, Senegal, crossing the Sahara Desert. It immediately attracted a motley crew of hard-core adventurers and racers who competed on a cosmopolitan range of machinery including Yamaha XT500s, Honda XR500s, and a couple of 800cc BMW Boxers. French privateer Jean-Claude Morellet raced in the first rally, showing the potential for a modified R80 machine in the desert and giving BMW the encouragement needed for a full factory effort. The company built a series of racebikes based on the new R80 G/S and would go on to win in 1981 with French rider Hubert Auriol, who also won in 1983. Auriol was joined by Belgian racer Gaston Rahier, who took victory in 1984 and 1985. Later success would come from the firm's single-cylinder machines: Richard Sainct won the 1999 Rally on an F650, and BMW took five wins between 1994 and 2002 in the women's trophy with riders Jutta Kleinschmidt and Andrea Mayer.

OPPOSITE TOP: Hubert Auriol in full flight during the 1984 Paris–Dakar Rally

OPPOSITE: Andrea Mayer bossing her F650 RR race special at the 2001 Paris–Dakar

ABOVE: Herbert Schek was the BMW engineer who designed and built the G/S racebike for Auriol—and he also built one for himself to compete too.

Barkbusters
Barkbusters
METZELER

10

2010s: Globalizing Performance

BMW Motorrad entered the 2010s in good shape. It had weathered the financial storms caused by the global financial crisis of 2008 and arguably coped better than the big Japanese bike firms. The R1200 GS was going from strength to strength as the class-defining machine of the burgeoning adventure-touring sector, while the smaller F800 GS, with almost as much capability as the big Boxer in a smaller, lighter, cheaper package, was gaining its own fan base. The other R-series Boxer and F-series parallel-twin models were doing well too, on a smaller scale, while the K1300 range was still strong, though sales of unlimited hyperbikes like the K1300 S were on the decline.

The biggest step forward came with the new S1000 RR. For perhaps the first time ever, BMW had given itself a shake and gone with the most straightforward, efficient design for the job in hand, despite it being the "conventional" choice. An inline-four, sixteen-valve, DOHC, water-cooled engine, mounted across the frame, with chain final drive, had been the mainstream Japanese layout since the mid-1980s. It powered all-out superbikes, tourers, commuters, and naked roadsters, was cheap and efficient to build, made strong, reliable horsepower, and was easy to package. BMW had *almost* cracked when designing the K100 twenty-five years earlier, going for an inline-four water-cooled layout, but then it compromised with the laid-down, shaft-drive packaging.

Still, better late than never, and the S1000 RR was an instant hit when it launched in mid-2009. Its conventional design didn't stop at the engine either; the chassis used a twin-spar aluminum frame, USD front forks, monoshock rear, and full fairing. It's hard to think of another BMW that had followed the crowd so closely before.

What prevented the new Bavarian superbike being a bit anticlimactic was its performance. The BMW engineers' time benchmarking the best in the liter superbike class had been well spent, and they'd clearly realized there'd be little point releasing a bike that didn't at least match the class leaders in terms of power and weight.

The S1000 RR's vital statistics were impressive. A claimed 193 bhp, with a wet weight of 204kg, comfortably matched competition like the Yamaha R1 (191 bhp/203kg) and topped the Suzuki GSX-R1000 (182 bhp/209kg), Kawasaki ZX-10R (188 bhp/208kg), and Honda Fireblade (175 bhp/200kg), on paper at least.

The author at the R1200 GS Rallye press launch in Portugal in 2017. ***Lel Pavey/BMW***

ABOVE: This stripped-down 2019 S1000 RR shows how compact and centralized the 207-bhp engine is. As with all the modern liter superbikes, the relationship between the engine's massive power, light weight, and diminutive size is hugely impressive.

OPPOSITE: For the first time, BMW had a genuine superbike without any "buts": no shaft drive, matching the best on power and weight, and at a very reasonable price too. It's no wonder the S1000 RR has been such a hit for BMW.

A New Superbike for the People

Best of all, the S1000 RR was priced to sell. The base bike cost £11,490 in the UK in 2010, midway between the £12,499 Yamaha R1 and £10,799 Honda Fireblade. By comparison, the base model of the Ducati 1198 was £13,295 (and the S version almost £17,000). The days of a flagship BMW motorcycle commanding an enormous cost premium were over, it seemed, and the market responded wildly. A whole new breed of riders, for whom BMW had never been an option, switched from Japanese sportsbikes to the new S1000 RR. Even in the hard-nosed world of British club racing, a small revolution took place, with riders dumping their R1s and GSX-Rs in favor of the German upstart. The pure performance gain outweighed the risk of swapping to a new, unknown machine, and as the BMW proved its reliability, it became the bike to have.

A New Range of Fours

BMW is well-practiced at getting the most out of a big investment like a new engine. And it wasn't long before the S1000 RR's success prompted two new variants. First came the 2014 S1000 R, a simple naked roadster conversion of the superbike. The engine was retuned to make 160 bhp rather than 193 bhp, a small nosecone fairing kept some of the wind blast off the rider, and it came with the

RR

2009–Present S1000 RR

It's hard to comprehend now what a turnaround the S1000 RR was for BMW. If you had told me in 1999 that within a decade, the Bavarian firm would best the Japanese in the 1,000cc sportsbike sector, I'd have questioned your sanity (or at least your sobriety). But the high-performance transformation begun by the K1200 S in 2004 was completed just five years later when the S1000 RR was unveiled at an event in Monza during the World Superbike round.

It was a genuine no-compromise sportsbike that could match the best of the sector in power, weight, and price. And it went even further in terms of technology. Advanced rider aids like a Race ABS system, "Slick" power mode for track use, and Dynamic Traction Control were beyond state of the art in 2009, while the eight-injector fuel injection, optional factory-fit quickshifter, and LCD instruments made much of the competition seem like analog competitors for a new digital age.

The chassis was perhaps less revolutionary, with frame, suspension, and brakes that could have featured on any other high-end liter superbike of the time. What marked it out was how well it worked, especially as a "first effort" in the superbike class. The light construction and chassis geometry gave superb handling on road and track and let the rider get the most from the incredible engine performance. Best of all, it was an easy bike to ride fast, with none of the fussy manners, peaky power delivery, or razor-edge handling of something like the MV Agusta F4 1000.

The styling is perhaps the only area where the original S1000 RR met criticism. The main fairing and headlamps used an asymmetrical design, giving an odd, robotic look instead of the conventionally attractive style of something like a Ducati 1198 Superbike. But it was the only area where BMW had stuck to its own slightly eccentric guns, and arguably it helped the S1000 RR define its own character, rather than being just another identikit mainstream superbike.

LEFT: Finally, BMW had done the sensible thing and made a simple, logical inline-four, sixteen-valve DOHC powerplant for its S1000 RR superbike. Applying the firm's massive performance know-how to a design shorn of compromises resulted in a superlative engine: compact, lightweight, reliable, and very powerful.

ABOVE: The first-generation S1000 RR had a curious asymmetrical look thanks to the odd-shaped headlights. The side fairing vents were also different on each side, as if the designers were making up for the RR's conventional engine and chassis setup with wildly unconventional styling.

OPPOSITE: The author riding a 2016 S1000 RR at the Qatar night launch of Michelin's Power RS sport tires. *Michelin*

The year 2019 saw a massive update of the S1000 RR, with a new ShiftCam engine and classier, symmetrical styling. BMW's superbike now had beauty that better matched its brawn.

S1000 Engine Development

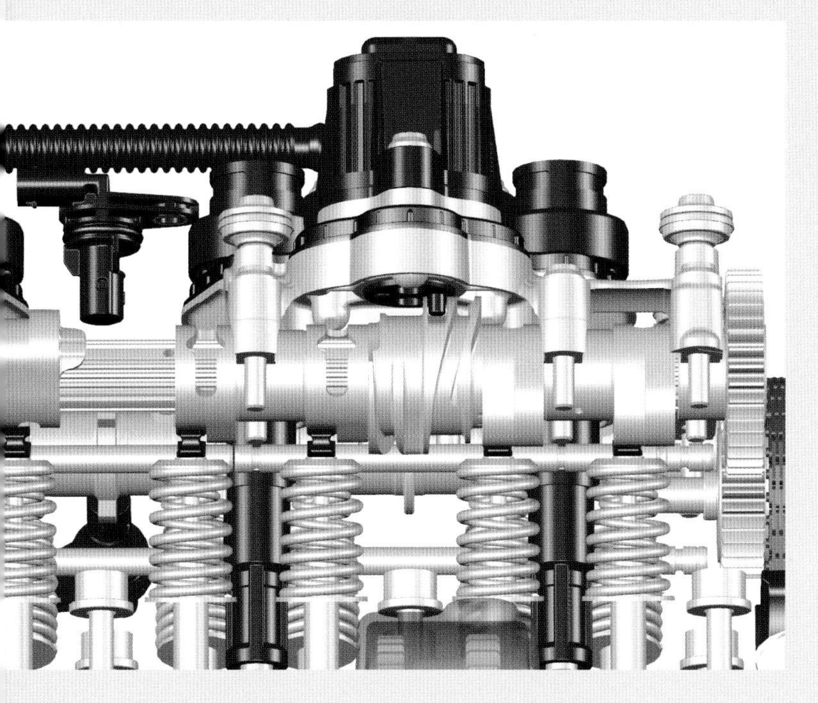

The original S1000 RR engine started as it meant to go on, and with a blank slate rather than updating a motor already in production, BMW engineers arguably had an advantage over their colleagues at other firms working with existing designs. The design was extreme from the start; a massive 80mm bore and 49.7mm stroke meant it had the most over-square cylinder architecture on the market, matching the layouts used in MotoGP engines. That allowed large titanium valves—33.5mm inlets and 27.2mm exhaust—which then gave superb air flow potential at high engine speeds. The cylinder head used finger-rocker valve actuation to permit a super-narrow valve angle for a compact combustion chamber design. The compression ratio was a heady 13:1, and the rev limit a wild 14,200 rpm. A 2012 midlife update to the intake and exhaust systems improved rideability but without any peak power increase.

For 2014, the S1000 engine was retuned for use in the S1000 R and S1000 XR, with different cams and revised cylinder head ports, plus changes to the engine management system, that together reduced peak power to "just" 160 bhp, with stronger midrange to suit.

The next update for the RR engine came for 2015, when a host of tweaks to the top end—cylinder head, exhaust, intake—saw an extra 6-bhp peak power, up to 199 bhp, plus more midrange torque. Then 2016 brought a new EU4 emissions-compliant tune for the engine, but it was the 2019 S1000 RR that saw a completely new engine design. The engine was totally overhauled and gained a variable-valve system called ShiftCam. This alters the intake valve timing and lift, thanks to a camshaft with concentric sliding sections. The outer sections have different cam profiles, one suitable for low rpm and one for high rpm. A sliding pin, controlled by the engine ECU, engages at 9,000 rpm and shifts the appropriate cam profile into place, altering the valve operation to suit engine speed. Power was now an incredible 207 bhp at 13,500 rpm, peak torque was 84 ft-lb at 10,500 rpm, and the motor was also 4kg lighter. The new RR engine was also introduced in a lower state of tune on the S1000 XR and S1000 R in 2020 and 2021, without the ShiftCam system (saving an extra 1kg on the motor).

The latest incarnation of the S1000 RR motor is in the M1000 RR, a limited edition "homologation" special designed to qualify for superbike racing. It has even more extreme performance, with two-ring forged pistons, titanium con rods, a redline of 15,100 rpm, and peak power output of 212 bhp.

ABOVE: For 2019, the S1000 RR gained a variable valve setup called ShiftCam. ECU-controlled solenoids engage pins in a sliding track, moving the outer section of the concentrically split camshaft along its axis. This illustration shows the pin and track and the splines locating the inner and outer camshaft sections.

BELOW: The 2020 S1000 XR engine was based on the 2019 RR engine, but, perhaps surprisingly, BMW dropped the Shiftcam setup and used conventional valve operation.

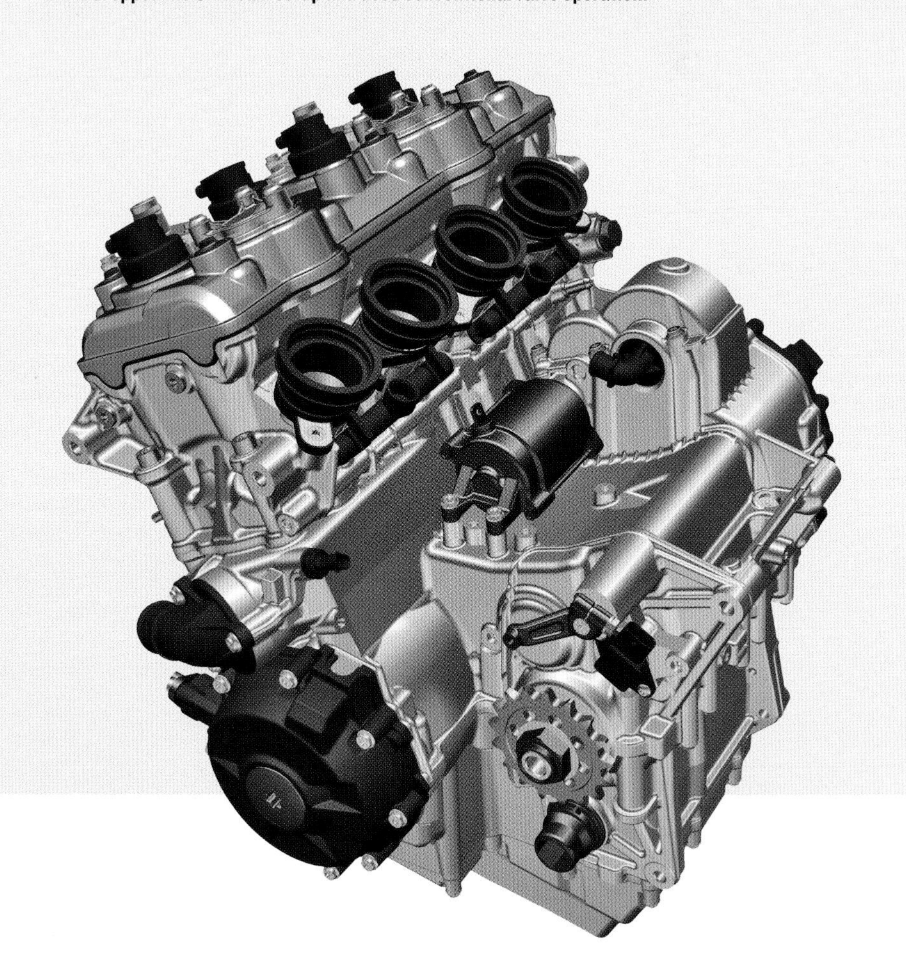

same high-end chassis components and electronic package as the RR. Like its sporty sibling, the S1000 R went straight to the top of its class, in this case the "super naked" sector that included the Triumph Speed Triple, Ducati Monster 1200, KTM Super Duke, and Aprilia Tuono V4.

The next S1000 was a bit of a curious one from BMW. The 2015 S1000 XR put the superbike-derived engine and chassis from the R model into a crossover/dual sport package, including longer travel suspension, tall fairing, upright riding position, and plenty of GS-inspired styling. Not unlike the Ducati Multistrada 1200, it aimed to combine the best of three worlds: superbike performance from a lightweight 160-bhp engine/chassis combination, mild off-road ability, and extensive touring capabilities. Like the Multistrada, it excelled at cross-country journeys, the suspension coping well on bumpy back roads, the engine providing the drive for sustained high speeds, and the comfortable riding position helping eat through motorway miles. It was also *a lot* of fun.

BELOW: The idea of a superbike-engined adventure touring bike seems wild but actually works really well. The S1000 XR makes a great high-performance distance machine, with sportsbike performance and handling, touring comfort, and slick style. You'd not want to take one off-road, mind.

OPPOSITE: The S1000 R offers almost all the performance and handling of the S1000 RR superbike in an easier, naked roadster package. The more upright riding position makes it a far better bike for urban and twisty back-road use.
James Wright

Arai
Arai

2012 HP4 and 2017 HP4 Race

Echoing the firm's HP2 range of the previous decade, BMW produced two special HP4 machines in the 2010s that were based on the S1000 RR. The first HP4 was a road bike and launched in late 2012, featuring the same 193-bhp engine output but with radically updated electronics and chassis package. It was the first bike with BMW's Dynamic Damping Control semiactive electronic suspension, and it also premiered a launch-control system, where an ECU cut torque as the front wheel lifted from a standstill, allowing full throttle launches without flipping the bike. Brembo monobloc brake calipers, lightweight forged wheels, full titanium exhaust, and a lightweight battery rounded off the extreme spec. It cost around £17,000 in the UK, a £4,000 premium over the stock S1000 RR.

The HP4 Race was a much more extreme affair. Launched in 2017, it was a rather strange beast: not at all road legal and not eligible for most racing either, it was a super-expensive (£68,000), limited-edition track day toy for wealthy BMW collectors. It had a carbon-fiber frame, a WSBK-spec rear swingarm made by Suter, carbon wheels, Öhlins FGR300 gas forks, Brembo GP4-R nickel-plated race monobloc calipers, and 2D dash—essentially the best of everything, everywhere you look. It had a stunning 146kg dry weight figure and made 215 bhp. I was lucky enough to ride one at a UK press launch at the Almeria circuit in Spain, and in terms of its performance, it easily ranked alongside full MotoGP and WSBK bikes I've ridden.

ABOVE: The HP4 was a limited factory-built special S1000 RR with BMW's most advanced chassis and engine tech and a cutting-edge electronics package.

OPPOSITE TOP AND BOTTOM: The track-only HP4 Race is the most extreme BMW ever built, with carbon frame and wheels, a factory race-tuned engine, and chassis kit that matches or tops what you see on World Superbike racing machines. Riding it at the UK press riding launch in Almeria was a breathtaking experience. *James Wright*

TOP: The first time the engine was displayed was in this naked roadster concept bike. Sadly (or luckily, depending on your point of view), this unfaired monster remains a concept: the K1600 inline-six engine has only been used in heavyweight tourers so far.

ABOVE: BMW showed off its inline-six engine before revealing what it would power.

The Joy of Six

Once again, in hindsight, the sheer work rate at BMW around 2010 was impressive. The S1000 was an enormous project, and you could see the R&D department being content with its success. But an even more audacious plan was afoot—the development of an all-new six-cylinder flagship range.

The first signs came, as they often do from BMW, in the form of a concept bike. The Concept 6, shown off at the international bike shows in late 2009, was a bruising powerbike, a naked muscle machine with a mighty straight-six engine at the center of the design. The following year, the K1600 GT and GTL launched with the powerplant as more conventional ultra-touring machines.

BMW has a long and successful association with six-cylinder engines in its car division. But for motorcycles, anything more than four cylinders has been an outlier. There were a few niche six-cylinder applications in the 1970s from Italian firm Benelli's Sei, the Honda CBX 1000, and Kawasaki's Z1300. All used in-line engines, and the excessive width and mass added compromises that outweighed any power or smoothness advantages.

For the new K1600s, however, BMW had worked hard to counteract the fundamental downsides of a six. The new DOHC, twenty-four-valve, 1,649cc engine was remarkably narrow, just 560mm wide (67mm wider than the K1300 unit), and the narrowest six ever used in a production machine. Despite that, it maintained a short-stroke bore-and-stroke layout rather than the longer-stroke designs used before. It only made 160 bhp, a figure easily bested by the far smaller S1000 RR motor. But it had a massive flat torque curve, with 175Nm or 129 ft-lb at 5,000 rpm, and 70 percent of that was available from just 1,500 rpm. The smoothness of a straight-six, with its perfect primary and secondary balance, is also unmatchable by a light, frenetic, four-cylinder superbike motor.

The new K1600s put BMW back on top in another key sector: the unlimited touring class. Where Honda's GL1800 Gold Wing could once sniff at the likes of the K1300 GT and the elderly K1200 LT, it had now met its match. The GT was the sportier of the range, with the larger GTL luxury variant offering extra comfort, integrated top box, and taller windshield. Several years later, BMW launched the even larger Grand America (2018) and low-slung B "bagger" (2016) variants aimed at the vast US touring bike market.

The K1600 GTL replaced the elderly **K1200 LT** as BMW's no-compromise ultra-touring machine. If you can handle the enormous size and 348kg wet mass, there's no more luxurious way to see the world on two wheels.

2011–Present K1600 GT/GTL

BMW's flagship luxury touring range marque met its current apotheosis with the 2011 K1600, a superlative distance machine based on the incredible straight-six engine, featuring a high-performance aluminum-framed chassis, Duolever front suspension, Paralever rear, an enormous protective fairing, and a host of high-tech rider aids and luxury gizmos. The 102.6kg engine block was laid down in similar fashion to the K1200 S, allowing the frame to pass over the top and keeping weight low on the bike.

Rider aids included the now-standard BMW fitments of ABS, traction control, and rider power modes, as well as the second-generation ESA II electronic suspension adjustment, an adaptive cornering headlight, cruise control, and a huge 5.7-inch TFT LCD color dashboard. Luxury touring options included GPS navigation and an audio system built into the dash, central locking, heated grips and seat, chrome trims, and much more.

For 2017, the K1600 range received a midlife face-lift that upgraded the rider aids to the latest versions and also added a new reverse assist, which used the starter motor to drive the 350kg+ bike backward, helping with parking maneuvers. In 2021, the K1600s got a new Euro5 emissions-compliant engine, Dynamic ESA semi-active suspension, and a 10.25-inch LCD dashboard, keeping the whole range at the top of the touring class.

One of the K1600's high-tech features was an automatic SOS system, which sensed a crash, and automatically called the emergency services for help.

Still Covering the Day Job

One danger of overreaching yourself with new, exciting projects is that you take your eye off the ball in your main bread-and-butter job. And you'd maybe excuse BMW for losing focus on its massively important Boxer range. The R1200 GS and R1200 GS Adventure were essentially unchallenged for most of the 2000s, and a series of minor updates had seemed enough to keep the huge sales figures coming. But the competition hadn't been totally asleep, and bikes like the Triumph Tiger 1200 and KTM Adventure 1190 were creeping up on the GS's hegemony.

Meanwhile, ever-tightening European engine emissions regulations were making life tougher for the oil- and air-cooled Boxer engine. The 1990s-designed powerplant had made massive improvements in power and sophistication, and the 1,170cc 1200 version was a superb unit. It had gained DOHC heads for 2009, with a cunning valve layout where the horizontally located camshafts operated one intake and one exhaust valve each. That kept the heads compact while sticking with the traditional Boxer horizontal intake and exhaust flow.

At first glance, it's not at all clear that the 2013 R1200 GS used water cooling. Since the water only cools the cylinder heads, the cylinders themselves retain their air-cooling fins, and the small radiators are well concealed. It was still down on peak power, but the overall package remained unbeatable.

RIGHT: **This cutaway diagram shows the unusual camshaft layout on the 2010 DOHC Boxer. Each shaft has one inlet and one exhaust cam nose.**

BELOW: **The 2013 engine's vertical gas flow into and out of the engine is clear in this graphic.**

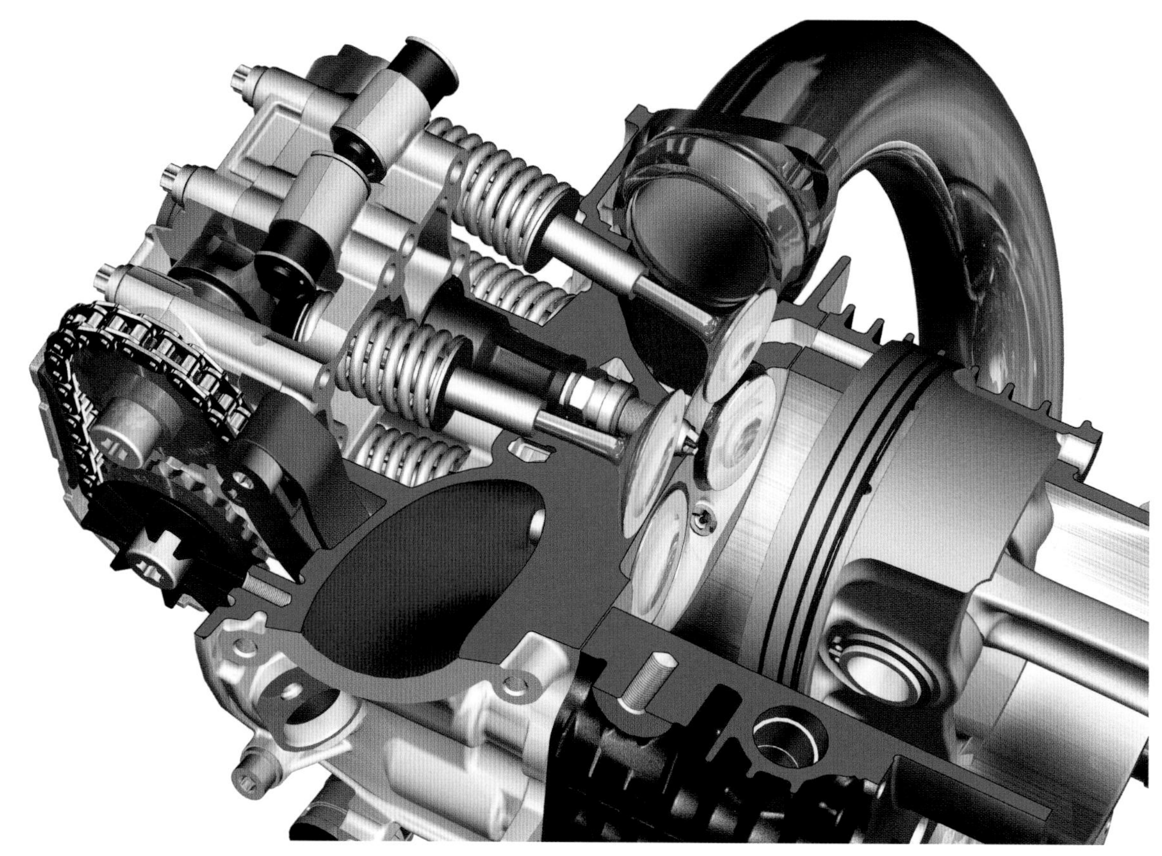

But with Euro4 emissions regs kicking in and more powerful competition appearing, the old oil-cooled motor needed an upgrade. That came for the 2013 model year; the all-new 1200 GS featured water-cooling and a revised intake/exhaust setup. The water circuit was only applied to the cylinder heads, and the inherent cooling advantages of the Boxer layout meant the cylinders could stick with oil and air to keep temperatures down. Careful design meant the water was only used where needed, reducing the size of the radiators and pipework required.

More fundamental was the change to a "vertical" intake and exhaust system. Since 1923, BMW Boxers had the intake in front of the rider's feet, with the exhaust ports pointing forward. Now the intake port was above the cylinder head, and the exhaust came straight out of the bottom, pointing at the ground, and the double overhead camshafts each operated only intake and exhaust valves.

The engine upgrades came alongside a continual ramping-up of equipment and technology levels. Systems formerly restricted to luxury tourers appeared on this hardened off-roader, including color LCD dashboards, cruise control, a multi-controller wheel on the switchgear, a Bluetooth-linked media and navigation system, and heated seats and grips. Meanwhile, the rider aids advanced in parallel—electronic semi-active suspension, cornering ABS and traction control, up and down quickshifter—you name it, it was available for the R1200 GS/Adventure.

Looking after the Other Twins

What about the other twin-cylinder BMWs, the F-series machinery? That range had grown in size, with the two sporty-touring F800 S and ST models, the F800 GS, and the confusing F650 GS twin (which had the same name as the F650 single and was actually an 800), plus the F800 R roadster. The R, S, and ST gradually declined in popularity

ABOVE: The 2016 F800 R saw a new ride-by-wire fuel injection setup and a styling face-lift.

LEFT: Replacing the older S and ST models, the F800 GT provided a middleweight option for BMW touring bike fans.

ABOVE: **The F900 XR took the styling of the S1000 XR to BMW's upper-middleweight parallel-twin range. A solid road touring package plus a larger-capacity version of the F850 GS engine makes a very decent long-distance commuter and tourer.** ***John Goodman***

OPPOSITE: **The F850 GS looked more like the bigger R1200 GS than ever, with extensive accessory options and hard-core styling. The weight and size was creeping up too.**

through the early 2010s, with the GS variants holding the fort in terms of F-series sales. A new F700 GS replaced the 650 in 2012, reducing confusion slightly, but it remained a detuned, lower-spec version of the 800. An F800 GT also appeared in 2012, replacing the S and ST in the sport-touring segment. For 2016, the F800 R and GT were updated with new ride-by-wire throttle, EU4-compliant engines, and a minor face-lift.

For 2018, BMW totally revised the F range, with new F750 GS and F850 GS models. They featured an all-new engine, with a conventional balancer shaft setup replacing the old rocking-beam system, and a 270-degree firing order. The 850 now made 95 bhp, and the 750 made 77 bhp, both using the same 853cc displacement. There was also a larger selection of options for the middleweight adventure range, including semi-active electronic suspension adjustment, keyless ignition, ABS Pro, LED lighting, and much more. Later in 2018, the F850 GS Adventure was released, with the usual Adventure upgrades: a larger 23-liter fuel tank, a tougher welded sheet steel frame, crash bars and rack, and longer-travel suspension.

An Even Bigger F

The latest update to the F series came in 2019 with the launch of the F900 R and F900 XR. The 850 GS engine was tweaked for more power (up to 105 bhp) from a larger 895cc capacity and bolted into a street-focused chassis setup. The designs aped the larger S1000 R and XR models, with the same naked roadster and road-biased adventure tourer feel, as well as similar technology and equipment options.

F850
BMW
GS
BMW

G 310
BMW
BMW

PREVIOUS PAGES: The 2015 G310 R was the smallest-capacity BMW bike since the 247cc R27 from 1966. Built in partnership with the giant Indian bike maker TVS, it marked a new level of globalization for BMW Motorrad. It was also a very capable entry-level roadster.

RIGHT: Producing a GS version of the G310 single was an obvious move by BMW to gain a new contender in the burgeoning mini-adventure bike sector.

Single Life

BMW's single-cylinder range had looked in danger of disappearing altogether in the mid-2000s, with the F650 twin seemingly set to replace the old F650 single. But the demand for a cheaper, lighter entry-level BMW remained strong, and the single returned as the G650 enduro in 2010. And that G moniker, first seen on the 2008 G450 X off-roader, would be adopted as the name for BMW's one-cylinder range.

The year 2015 brought a brand-new single and the smallest-capacity production BMW ever: the G310 R roadster. It was built in partnership with Indian firm TVS as a new super-lightweight gateway into BMW ownership. Its 34-bhp power output came from an innovative 313cc engine design, featuring a "reverse" cylinder head that had the intake at the front of a backward-tilted cylinder. The exhaust exited from the rear of the head, passing down through the rear swingarm. The chassis was simple stuff: a steel-tube frame, USD forks, and a dual-sided aluminum rear swingarm. A dual-channel ABS system and LCD dash added a touch of high tech, but the G310 R was a pretty straightforward piece of kit.

The R was joined by a mini-adventure version in 2017: the G310 GS. This took the same basic engine and frame and added longer-travel suspension, GS-style bodywork, and dirt-friendly wheel sizes. The sub-500cc adventure category expanded through the 2010s, as riders rebelled against the ever-increasing girth of the larger adventure tourers, and the G310 GS was a sound option when compared with the likes of the Honda CRF300, Kawasaki Versys-X 300, and Suzuki V-Strom 250.

Going Back to the Future

With the ever-increasing complexity and sophistication of the "mainstream" water-cooled R1200 Boxer range, you'd think that the old Oilhead Boxer engine would have gone the way of all flesh. For 2014, the R1200 GS, Adventure, and RT all had the new motor, and the R1200 RS and R1200 R would follow over the next twelve months.

BMW had other plans for the Oilhead though. There had been an explosion of interest across the bike industry in retro or "modern classic" designs. Firms like Triumph had long been selling brand-new bikes with styling from thirty, forty, even fifty years ago. Ducati had experimented with retro-styled sport classics, and the market in used, restored bikes from the 1970s and 1980s was booming. With the typical European and American bike rider now aged between forty and sixty years old, the appetite for nostalgia had never been so big.

And BMW, as you've seen in this very tome, has a lot of heritage to be nostalgic about. What better way to use the "old" Boxer motor than in an "old" retro roadster design?

A concept machine appeared in 2013 at the Concorso d'Eleganza Villa d'Este on Lake Como in Italy. Dubbed the Concept Ninety, it was part of BMW Motorrad's ninetieth-anniversary celebrations and a clear homage to the mighty R90 S that had transformed BMW's reputation in the early 1970s.

The Munich firm had gone to America, home of so much R90 S success, for the concept machine and signed up legendary customizer and racer Roland Sands for the job. Based on the traditional Oilhead powertrain, it added MotoGP-spec running gear, Dunlop slicks, and a gorgeous, modern take on the R90 S style.

From Concept to R nineT

The Concept Ninety was, of course, a prequel to the launch of the production machine, dubbed the R nineT. The name is a little clumsy from a grammatical point of view, but it clearly pays homage to the original R90 in a modern form, which is just what the bike itself did. There was nothing earth-shattering about the design: the fundamental design was not unlike an R1200 R Classic, with the Telelever swapped for the premium USD forks and radial brakes from the S1000 R. The styling was less extreme than Roland Sands's Concept Ninety machine, sadly, but was an attractive blend of retro and modern. The key to the R nineT design, though, was its flexibility. The designers had purposefully built it to be easily customized, with an easily

Early design sketches show how the R nineT styling evolved.

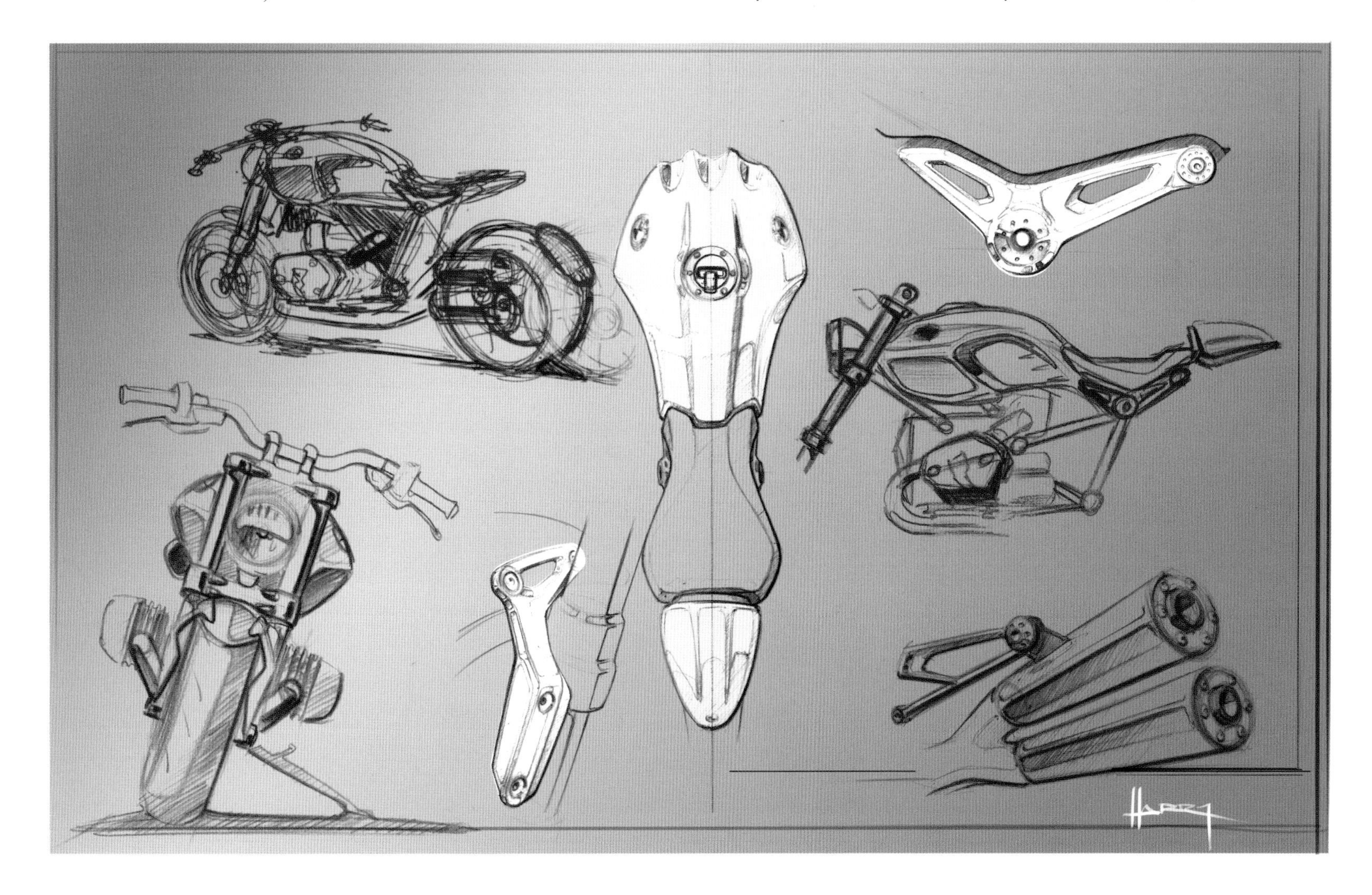

With the R nineT, BMW produced a high-quality roadster with good performance that was packed with sharp styling.

removable rear subframe and a massive preproduced range of aftermarket styling parts. The idea was you would buy your R nineT, choose from a menu of factory- or dealer-fit options to suit your own design, then take it away and bolt on even more fancy custom parts.

The new retro machine was a big hit and had come at exactly the right time. Ducati launched its modern retro Scrambler range the same year, and firms like Yamaha, Kawasaki, Honda, and Suzuki all followed with their own heritage lines over the next few years. BMW couldn't keep up with demand for the original R nineT and soon expanded the series into a full range. The R nineT was joined by the R nineT Scrambler, R nineT Pure, R nineT Racer, and R nineT Urban G/S, all using the older Oilhead engine and basic chassis, with varying levels of customization. The Scrambler added mild 1960s off-road styling, while the Pure was a cheaper entry-level model with lower-spec suspension. The Racer gained a natty half-fairing and extreme café racer riding position, and the Urban G/S took its cues from the original R80 G/S and put that into a slick city crosser.

More Capacity, More Tech for the R1200 Range

The next chapter in the mainstream Boxer range appeared in 2019 with the new R1250 GS and RT, both featuring a new ShiftCam variable valve system, larger capacity, and the most power ever from a BMW twin. The ShiftCam, also seen on the 2019 S1000 RR, operates on the intake camshaft and has two cam profiles, one for high rpm and one for low. The cams are mounted on a sliding concentric splined outer shaft and are moved axially by an ECU-controlled pin. When the pin engages in a cam track (similar to a gearbox selector drum design), the whole outer shaft moves along, switching from one profile to the other.

Having two separate valve timing and lift profiles allows much more efficient, cleaner power production.

The R1250 unit, now 1,254cc, put out 136 bhp (up from 125 bhp on the 1200) and 106 ft-lb of torque and was soon extended from the GS and RT to the R1250 R, GS Adventure, and RS models, replacing the 1200 engine across the board.

A Boxer Behemoth

As the 2010s ended, BMW surprised everyone once more with another all-new bike—the 2020 R18. It was powered by the biggest Boxer engine ever, featuring an air-cooled pushrod OHV design and a gigantic 1,802cc capacity. It came in a cruiser chassis with design cues from the 1930s R5 and R51 and was aimed at a sector that the company had neglected for nearly two decades—the big-bore cruiser.

This part of the market is dominated by Harley-Davidson, of course, and it requires a very specific mix of technology, style, and heritage to succeed, particularly in the United States. The R1200 C (BMW's last proper cruiser) didn't really hit the mark, but this new design looked like a real contender. Rather than repurposing an existing bike

The Racer version of the R nineT looks incredible, with its neat café-racer fairing, low-slung clip-on handlebars, and high pegs. The riding position is extreme as a result, and it is hard work on longer journeys.

as the R1200 C had done, the R18 was designed from the start to compete with the likes of Harley-Davidson; its "proper" air-cooled engine was built with style as much as performance in mind. It made around 90 bhp, with 111 ft-lb of torque (on par with the likes of a Harley big-twin) but had to carry a lot of mass too: the basic R18 weighed in at 345kg ready-to-ride. Under the old-tech exterior, the new motor was packed with technology. That's essential to meet emissions standards, but the ride-by-wire fuel injection also allowed modern rider aids like traction control, engine brake control, and three rider power modes dubbed Rock, Roll, and Rain. Keyless ignition, cruise control, and heated grips all continued the twenty-first-century take on a twentieth-century retro-cruiser.

BMW extended the R18 range through 2021, with a Classic model featuring soft luggage and windshield, the Transcontinental heavy tourer, and the R18 B bagger. All used the same basic engine and chassis, as well as retro-styled fairings, luggage, and audio systems by the legendary British Marshall amplifier company.

BELOW: **For 2021, BMW released a "40 Years of GS" paint scheme for the Urban G/S R nineT. The black and yellow paint scheme perfectly echoes a 1980s GS paint job.**

OPPOSITE: **The enormous R18 handles far better than its sheer weight would suggest. It gives a very modern, smooth, capable ride, thanks to the very latest engine management setup and electronic riding aids, belying its almost vintage classic BMW styling.**

Arai
R18

R18

The R18 range was expanded in 2021 with two new touring models, the Transcontinental and the B bagger, aimed at the American "heritage tourer" market. With performance and attitude to match Harley-Davidson's touring twins, BMW is hoping its extra levels of equipment can gain it a foothold in this lucrative US sector.

Roland Sands on the Concept 90 and the R nineT

The R nineT was such a departure for BMW and has been so successful that I wanted to speak to Roland Sands about it directly. Here's how a California-based customizer and racer became part of one of BMW's most successful bike ranges.

How did the Concept Ninety project begin?

Ola Stengard, who was the lead designer, is a friend of mine, and we hit it off a long time ago. We probably talked about the project for four years before it became a reality. Ola actually rode a prototype to my shop and dropped it off with the idea of us doing this collaboration. It was kind of like, "We're dropping this bike off—decide if you want to build something cool out of it." I took it up the street, did a few wheelies, and was impressed, and that was how my love affair with the Boxer started. We ended the evening with a massive burnout in my parking lot, and that was maybe the first burnout that bike did, but not the last.

What was the timeline? Did you work mostly in the US or travel to Munich? Was BMW Motorrad USA a big part of the arrangement?

The project was mostly in the US, but I did spend time in Munich with the design team working on the concept, the parts ideation, and the idea as a whole. The [accessory] parts were part of the deal from the beginning, and Stephan Schaller, who was the big boss at the time, was very gung-ho to get parts that BMW could sell attached to the project. Timeline was like six months; we had to get it finished for Villa Des in Italy on Lake Como. The launch of the bike was one of those dream memories, with an incredible ride up the coast of Lake Como on an RT. I want to go back.

What were your inspirations for the concept bike design? What were you aiming for?

BMW did the bodywork design, and it was up to us

TOP: Roland Sands in his workshop, finalizing the details on the Concept Ninety machine

BELOW: What better way to round off a shakedown test ride on a new bike than a mighty burnout? Many concept show bikes have nothing more than empty engine cases under their fake bodywork. But Sands insisted that the Concept Ninety would ride as well as it looked.

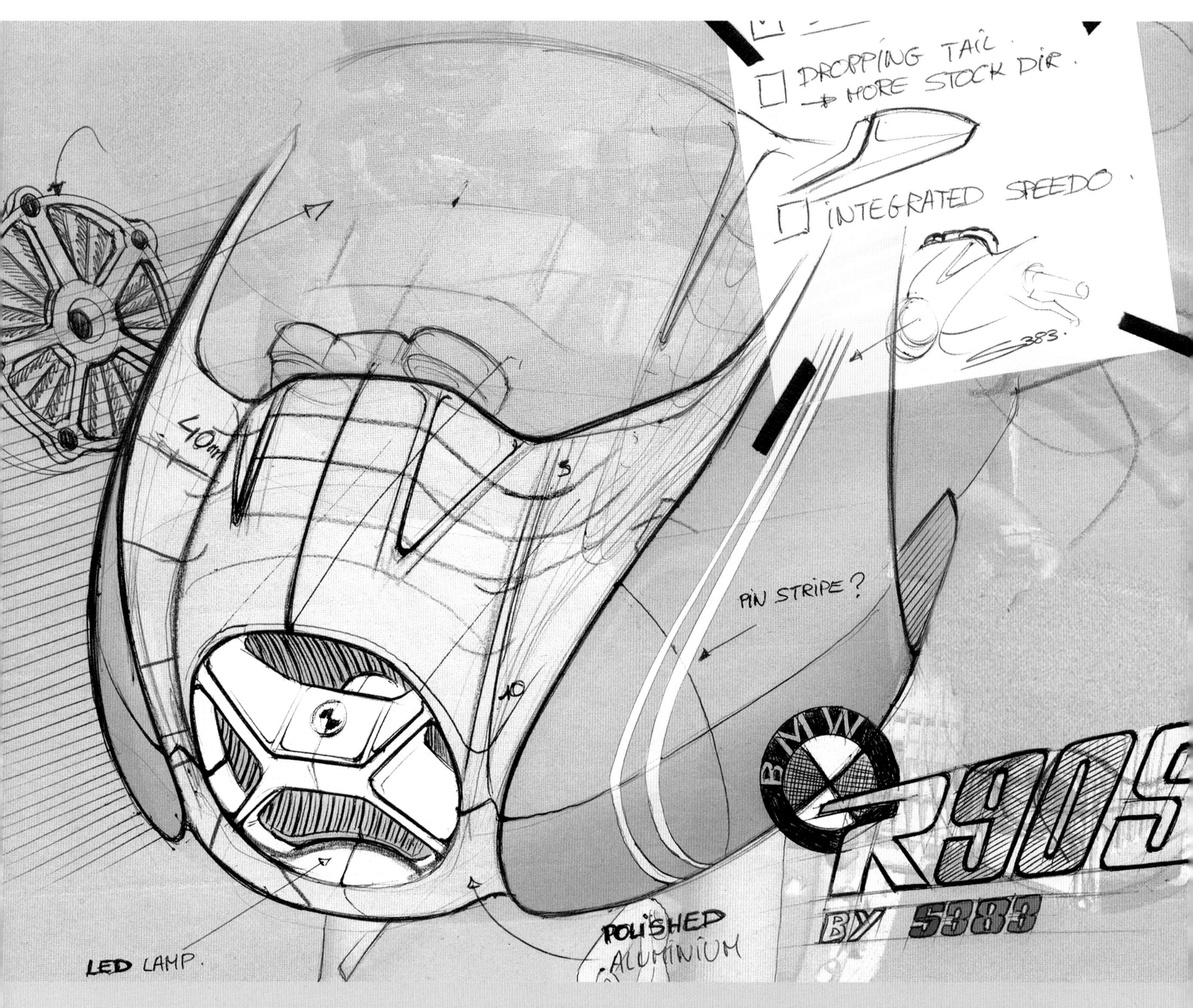

to execute, so of course we put our own spin on everything and designed twelve unique and new accessory parts to go with it. We for sure looked back at the original Reg Pridmore Daytona racer R90 S and upgraded suspension, brakes, and wheels. The bike has Dunlop road race slicks on it, so it was built to shred. When we did the filming of the Concept, the BMW guys were so freaked out because the bike was ultra top secret and I was dragging my knee on it at Willow Springs. They weren't used to concept bikes being ridden at all—normally they are static. But Ola pushed everyone to build a running prototype, and I suppose that's the main reason they wanted us to work on the bike.

ABOVE AND FOLLOWING PAGES: The influence of the 1970s R90 S on the R nineT is clear in these stunning color design sketches.

BMW
ÖHLINS
brembo
MICHELIN
MICHELIN
COMMANDER II

R 90 S
ÖHLINS
MICHELIN
COMMANDER II

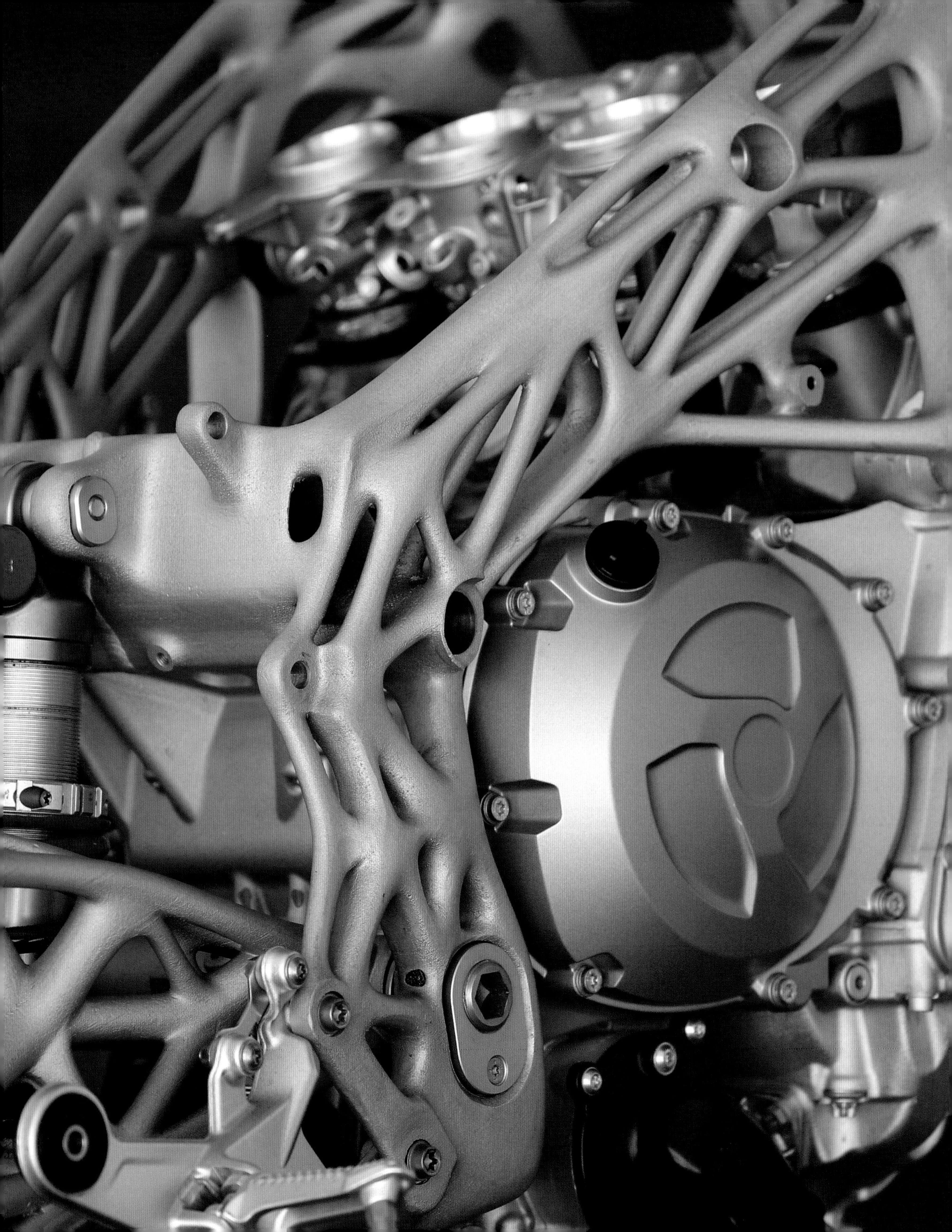

Do Androids Dream of Electric Bikes?

Toward the end of summer 2018, I received an invite to a BMW technical event. But the plane ticket didn't have Munich or Berlin as the destination; I was going to Marseille in the south of France. BMW has an R&D center in the suburb of Miramas, with a secure, private proving ground for cars and bikes. Its location on the French Riviera enjoys good weather all year round, allowing outdoor testing when Germany is shivering in ice and snow. The technical presentations covered some interesting areas: innovative carbon-fiber production methods as used on the HP4 Race and some of the testing methods used to prove the safety of carbon-fiber wheels and frames, as well as new 3D printing methods for metal parts. But the most interesting display came at the end of the day, when we were taken to a small asphalt apron outside a test building. Some technicians were busy at a computer screen, and we heard the unmistakable sound of a Boxer engine starting up and riding around the building. An R1200 GS appeared with a test rider on board and parked in front of us. The rider put down the kickstand, jumped off, and walked over to the technicians. Then the GS started up—by itself—blipped its throttle, and moved off, with no rider on board. The kickstand flicked up, and the seemingly possessed GS went through a series of maneuvers on the tarmac paddock area in front of us. It wasn't hanging about—this was no slow-speed demo—and the bike did a quick series of figure eights, swooping curves, and sharp turns all in the fairly tight space available. Then it came to a halt, popped out its kickstand, and parked itself, engine ticking over, seemingly poised to do whatever it wanted next.

The engineers took us through the massive amounts of kit on this self-riding GS: the top box and panniers were packed with computers, and there were GPS and inertial measurement units, cameras, laser range finders, and sensor packages all over the bike. There was also a robotic actuator on the steering, plus computer-controlled clutch, gear change, throttle, and kickstand mechanisms. It looked to be a remarkably simple package, with no massive gyroscopes spinning to keep the bike

The frame of the future? A 3D-printed metal unit at BMW's Marseilles R&D center.

balanced upright or funny extending steering heads as seen on the self-riding motorcycle prototype that Honda showed off in 2017.

A Pointless Invention?

BMW was working hard on producing a motorcycle that could ride itself then. But what would be the point of a self-riding motorcycle? A self-driving car, delivery truck, drone aircraft, or taxi all makes sense, but the point of riding a motorbike is, generally, to enjoy the dynamic experience of being in control. Who but the most eccentric of commuters would sit astride a self-riding motorcycle to get from A to B without any involvement?

The answer is a little more subtle. BMW (alongside many other firms) was researching how a computer

ABOVE: BMW's R&D center showed off this additive-manufacture 3D-printed S1000 RR frame and swingarm. The organic shape of the structure offers optimal stiffness and strength with minimal mass—in theory, at least. It's some way off production at the moment.

LEFT: Light and stiff, carbon fiber is an ideal material for components like wheels. But the critical safety nature of these parts means that very high testing and quality standards are needed throughout production.

could step in and take over the riding function of a bike. Not to ride the bike in "normal" times but so it could assume control from the rider in an instant to avoid a crash. Rather like how an ABS computer takes charge of the brakes when you grab a big handful on some wet leaves, the self-riding technology would form a sort of "anti-crash system," or ACS. Linked with the "smart highway" networks that are coming across the globe and communicating with self-driving cars and buses, roadside sensors, and traffic signals, the brain of this ACS is constantly watching for trouble, just as your ABS computer is constantly looking for a locked front wheel. The ACS might "see" that a truck is about to pull out of a hidden junction or that a tractor has broken down just over the blind summit ahead. It might know that there's a patch of ice around the next bend or realize that you're riding 20 mph too fast to make the corner coming up. Prepared for what's about to happen, it can intervene, reduce speed, start steering around an obstacle, lean over a bit more, or hit the brakes, preventing the crash it predicted. It would step in the same way your ABS and traction control ECUs do, but preventing a much wider range of mishaps.

Science fiction? Maybe. But it's the same way of thinking that lies behind the technical advances in rider technology that BMW's been developing since the early 1980s. Back then, ABS seemed like witchcraft, and many riders reckoned they would rather rely on their own skills than some damn

This car/bike demo showed how future cars and bikes would communicate constantly over 5G networks to avoid collisions at junctions and other crashes.

R1200
GS
BMW

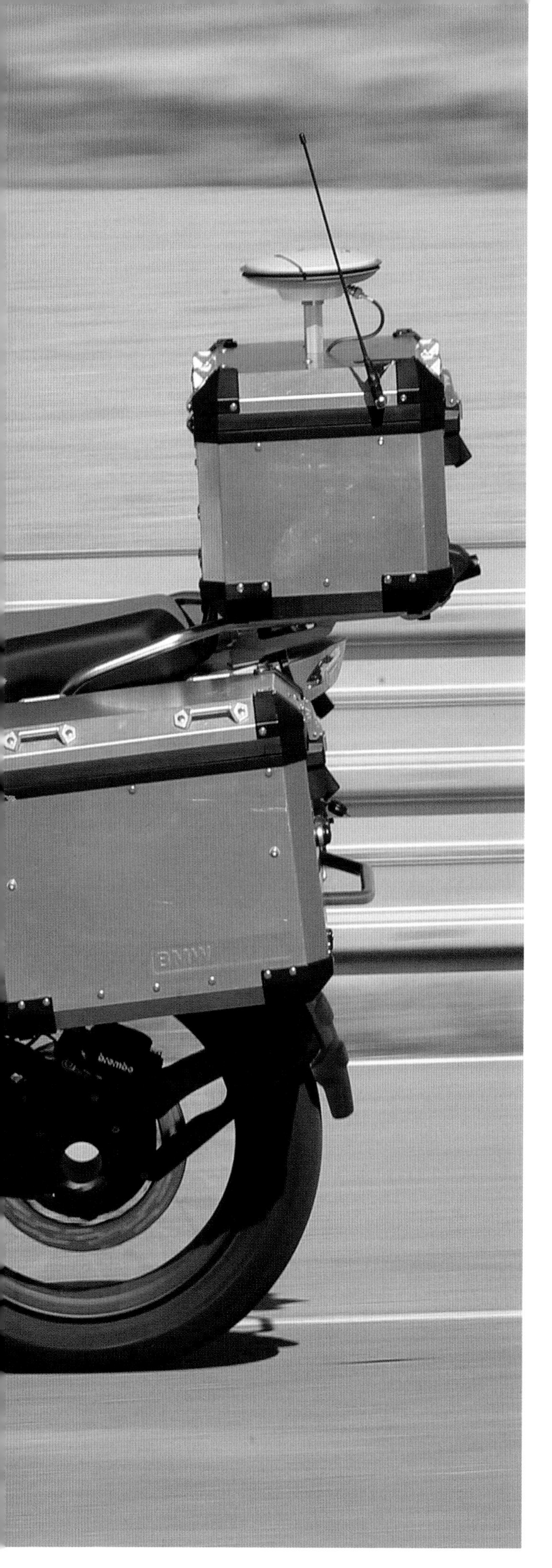

computer under the seat. And, indeed, early ABS setups weren't as effective as a very skilled rider when it came to stopping hard, especially on the track. As the technology has improved over the past four decades, however, current ABS systems perform far better and incorporate technologies like IMU-based cornering algorithms that put them beyond human performance. Added to the latest DTC (dynamic traction control) setups, you have a safety system that can help avoid many of the "single-vehicle loss of control" accident statistics that plague motorcycling. Extending a bike's onboard systems to prevent other types of accidents seems like simple logic.

ABOVE AND OPPOSITE: **With a topbox and panniers full of computers, GPS, sensors, and communication network gear, this R1200 GS can ride itself.**

Better Living through Technology

When the first-ever BMW motorcycle was released in 1923, the technology of the time was just about up to providing a working internal combustion engine with ignition, fueling, and valve control. Gradually, luxuries like electric lighting began to appear, but as with most of the motorcycle (and car) industry, BMW bikes were fairly austere machines right up until the 1970s. Then the R90 S appeared with its small nosecone fairing, followed by the R100 RS with a larger frame-mounted fairing, and rider comfort and convenience became a real selling point. Hard luggage and more comprehensive dashboard instrumentation

followed, and then in the 1980s and 1990s, genuine luxury began to arrive. Sound systems, heated grips and seats, trip computers, and intercoms all became part of the riding experience, putting BMW owners on par with luxury car drivers in terms of the kit available to ease the burden on long journeys.

The twenty-first century saw BMW pull away even further. GPS navigation systems become widespread, and the firm began a long association with Garmin, using BMW-branded navigation units mounted in secure brackets. Later, they would integrate with the dashboards, which themselves moved to larger, full-color LCD designs. Switchgear evolved too, with extra menu selection buttons and a large, ergonomic controller wheel on the left-hand bar to allow easy navigation through menu options and bike setup pages.

Safety and Comfort

Some tech kit boosts both comfort/convenience and safety. BMW's Electronic Suspension Adjustment and semi-active Dynamic ESA setups increase both rider comfort and chassis performance, boosting stability and improving grip while giving a more luxurious ride. Integrated tire pressure monitoring saves getting your hands dirty checking pressures with an old-school gauge, and it gives early warning of dangerous slow punctures or underinflated tires. Linking the bike to your phone adds a Bluetooth telematics function so you can take calls, listen to music, and integrate navigation features with Apple and Android smartphones via BMW's own app. But it also adds an SOS function, so your bike can automatically call for help if you have a crash and are unconscious or injured. And the Adaptive Cruise Control launched on the 2021 R1250 RT has a radar unit that measures the distance to vehicles in front, keeping a safe distance automatically.

Electric Epilogue

As BMW Motorrad passes its one-hundredth anniversary, it's heading into an uncertain future. The basis of its entire output—the gasoline-powered internal combustion engine—looks to be heading the way of the horse-based transport systems it supplanted in the early twentieth century. Where the cities of London, New York, Tokyo, and Munich

ABOVE: BMW's controller wheel is intuitive and easy to use when navigating through a range of bike system setups. It spins forward and back to scroll through menus and clicks in to select options.

RIGHT: The amount of information available to the latest BMW riders is mind-bending. From air temperature to fuel consumption, range remaining, tire pressures, battery voltage, mobile phone signal quality, electronic suspension settings, engine power modes, ABS function, sound system, navigation, SMS messages—you name it, a BMW dash can probably display it.

were being choked by horse manure a century ago, they're now having to deal with dangerous emissions of a less obvious, more threatening kind. Burning fossil fuels, and the associated increase in atmospheric carbon dioxide levels, is warming the planet, with potentially dangerous results. And while motorcycles are a very small part of the problem, they have to be a part of the solution, moving away from gasoline to other forms of power.

At the moment, that looks to be electrical power, though it's not straightforward on bikes. In the automotive world, battery-powered electric motors have become widespread as a replacement for gas and diesel powertrains. But cars have plenty of space for large battery packs and can easily accommodate the weight. Bikes are much more limited on both fronts—hiding a 300kg battery pack the size of two or three large suitcases is simple enough in a family hatchback missing its gas tank but is very tricky on a motorcycle. Range and power output are directly linked to the battery pack size and weight, so to have a bike that matches gas-engine power outputs and range requires a big, heavy power pack. And while charging technology is improving daily, driven by the requirements of the car world, it's still much slower than filling a tank with fuel.

Baby Steps

BMW has dabbled with electric drive on two wheels with two production models so far, both scooters aimed at urban commuting. The C Evolution was on sale from 2014 to 2020 in limited numbers and was the first electric two-wheeler from a mainstream firm. It made a solid 47 bhp and had up to 99-mile range, using battery technology from the firm's i3 electric car. Speed was limited to 80 mph, giving it the performance of a 300cc-class scooter, though it was much heavier, at 265kg, and had conventional maxi-scooter styling.

In 2021, the CE-04 introduced a much more radical, futuristic design. Smaller and lighter, with unconventional touches like the side-entry underseat storage, it had slightly less power (42 bhp peak) and range (80 miles) but added fast-charging technology and a lower price.

Both the CE scooters are excellent vehicles, and they are loved by owners willing to adapt to the compromises of an electric urban bike. They're a long

ABOVE: **BMW's latest touring aid comes via radar. A sensor on the front of the latest R1250 RT measures the distance to vehicles in front and maintains a safe distance while the cruise control is engaged.**

BELOW: **Under the SOS cover is an emergency button that can be used to make an emergency call via a smartphone link. The system can also make automatic calls if a crash is detected.**

Ride-by-Wire Throttle

A vital part of the safety systems used on bikes is an accurate method of controlling the engine output. Anti-wheelie control, traction control, engine brake control, plus performance features like launch control and slide control, all need precise computer input into what the engine is doing. Early traction control systems used fairly primitive methods to reduce power when the rear wheel began to spin; they would cut the ignition spark or switch off the fuel injectors. The intervention was abrupt, slow to recover, and like early ABS systems, unsettling for bike and rider.

Ride-by-wire (RBW) engine management allows much finer control. Here, rather than a simple steel Bowden cable opening a throttle valve into the engine, the twist grip merely sends a signal to an ECU, which then opens the throttle valves via an electric motor. That ECU is also running the traction, slide, anti-wheelie, engine braking, and other safety routines at the same time. If it thinks the bike is getting into trouble or about to do something unwanted, it can close the throttle itself, reducing the torque from the engine and avoiding a crash.

The 2019 S1000 RR also used split throttle bodies: there are two throttle valve actuators, each controlling one pair of cylinders, which in theory allows finer traction control. At high lean angles, the ECU can keep a pair of throttles closed for an instant, feeding power in more gradually with the other pair. They're not used on the road bike, however, and are merely there for race homologation rules.

The 2019 S1000 RR uses the latest ride-by-wire fuel injection. It has computer-controlled butterfly valves operated by two stepper motors, seen here on each side of the throttle body unit, below the intake trumpet. Two throttle motors let race tuners split the left- and right-hand throttle pairs for even finer control.

LEFT: **Back in 2015, BMW showed off a prototype electric superbike, the E RR. It was based heavily on an S1000 RR chassis, with a battery pack and electric motor under the fairing. Few details were released at the time, and there's been no word since on any production application of the technology used.**

BELOW: **BMW's first production electric two-wheeler was the C Evolution scooter. This stripped-down shot shows how much space the battery pack takes up, even for a modest range.**

way from BMW's mainstream, though, and it's hard to see where an electric version of the R1250 GS or S1000 RR will come from in the short to medium term. But with governments around the world pledging to phase out gasoline engines in the 2030s, electric bikes seem to be coming whether we like it or not (though you might put an outside bet on niche technologies like hydrogen power or synthetic hydrocarbon fuels made using solar power to smash together hydrogen from water and carbon from CO2 in the atmosphere).

Add in the advanced safety protocols from connected highways and self-driving AI-powered vehicles, and you could foresee a very different type of personal two-wheeled transport. Small, electric, crash-proof motorcycles would be an amazing achievement in many ways. But for many, they would be the antithesis of what motorcycling is all about: the noise, the danger, the freedom.

But will the fun and excitement that has driven BMW's motorcycles for the past one hundred years still be there? If any firm can manage it, you'd have to bet on the Bavarian behemoth doing the business.

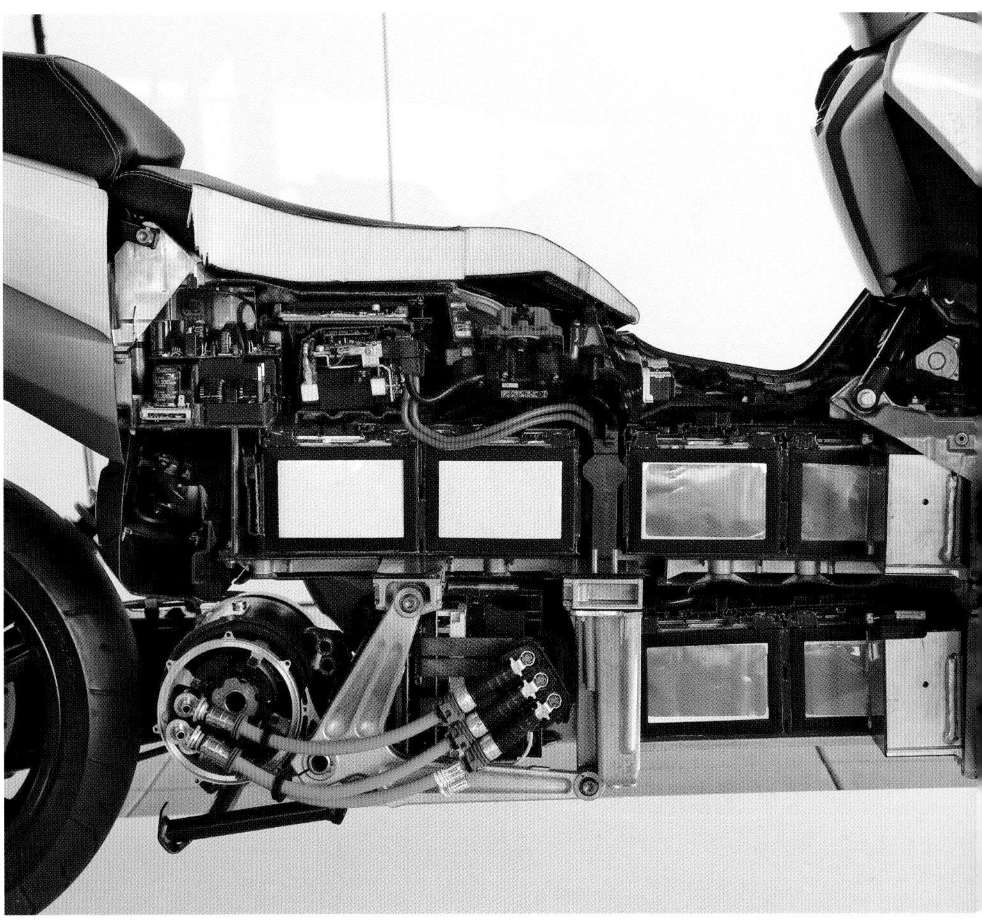

BMW
01
CE-04

The latest electric bike is another scooter, this time the avant-garde CE 04. It's very much an urban mobility solution rather than a traditional motorcycle design; it's hard to see traditional bike fans swapping their Boxers for a CE 04.

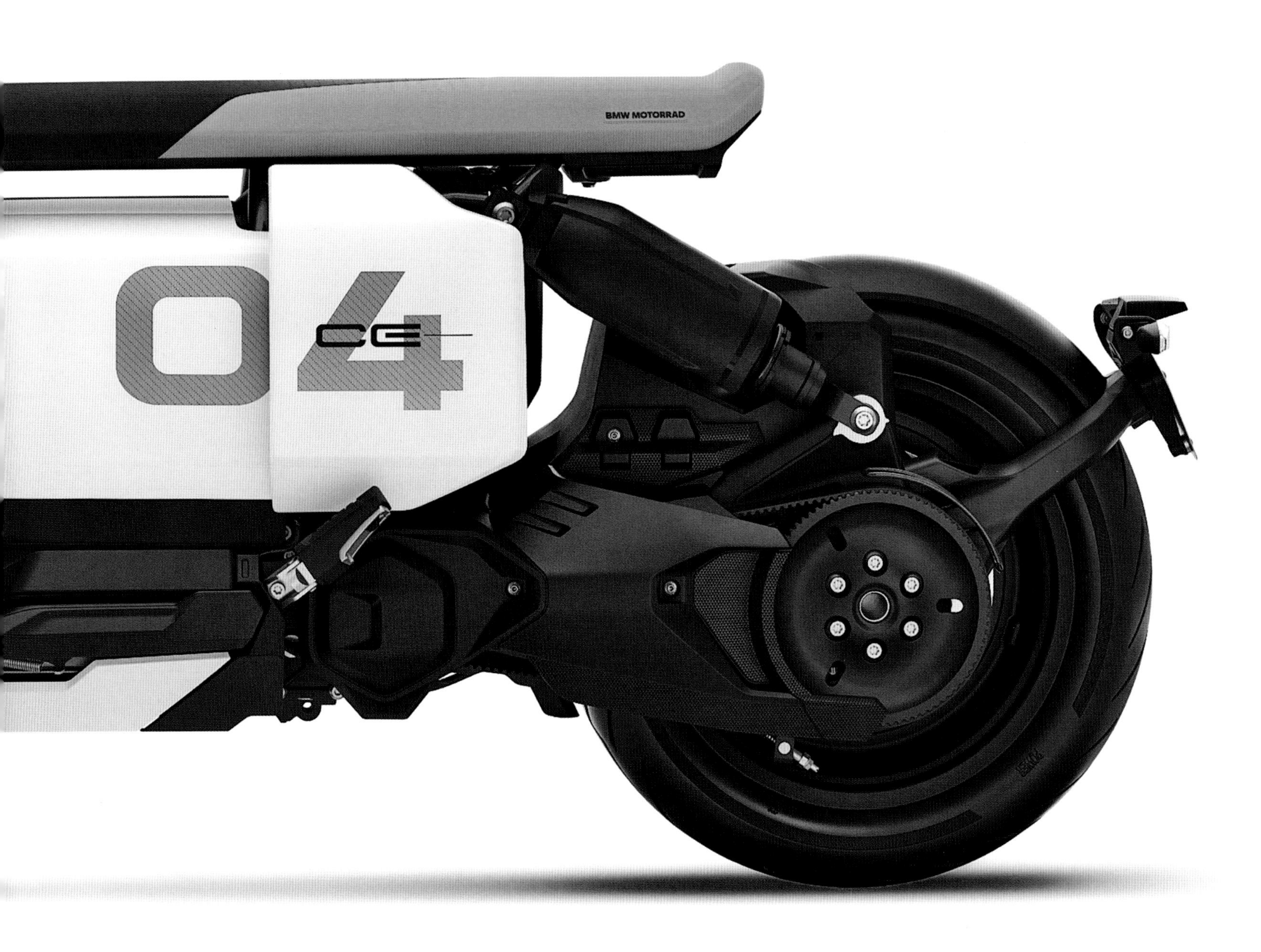

BMW Motorrad

Index

Acknowledgments

Enormous thanks to the good people at BMW who helped me with my research: Scott Grimsdall and Neil Allen in the UK, and Lutz-Michael Hahn and Ruth Standfuss in Germany.

Thanks also to the photographers who've taken great pictures of me riding various BMW press bikes over the years—James Wright, Jason Critchell, Lel Pavey, Gary Chapman, Joe Dick, John Noble, and John Goodman.

Finally, thanks to John Hogan for access to the *SuperBike* magazine digital archives.